PROTOCOLS FOR HIGH-EFFICIENCY WIRELESS NETWORKS

PROTOCOLS FOR HIGH-EFFICIENCY WIRELESS NETWORKS

by

Alessandro Andreadis
Giovanni Giambene

KLUWER ACADEMIC PUBLISHERS
Boston / Dordrecht / London

Distributors for North, Central and South America:
Kluwer Academic Publishers
101 Philip Drive
Assinippi Park
Norwell, Massachusetts 02061 USA
Telephone (781) 871-6600
Fax (781) 681-9045
E-Mail: kluwer@wkap.com

Distributors for all other countries:
Kluwer Academic Publishers Group
Post Office Box 322
3300 AH Dordrecht, THE NETHERLANDS
Telephone 31 78 6576 000
Fax 31 78 6576 254
E-Mail: services@wkap.nl

 Electronic Services < http://www.wkap.nl>

Library of Congress Cataloging-in-Publication Data

Protocols for High-Efficiency Wireless Networks
Alessandro Andreadis and Giovanni Giambene
ISBN 1-4020-7326-7

Printed on acid-free paper.

Printed in the United States of America

Acknowledgments:

The authors wish to thank Prof. Giuliano Benelli for his continuous help and encouragement.

Table of contents

Preface

Radio transmissions have opened new frontiers allowing the exchange of information with remote units. From the first applications of telegraphy and radio broadcast, wireless transmissions have obtained a great success with the widespread diffusion of mobile communications.

We live in the communication era, where any kind of information must be easy accessible to any user at any time. Mobile communication systems are the technical support that allows the realization of such concepts.

With the term mobile communications we embrace a set of technologies for radio transmissions, network protocols, mobile terminals and network elements.

The widespread diffusion of wireless communications is making national borders irrelevant in the design, delivery and billing of services, thus requiring international coordination of standardization efforts in order to evolve regional systems towards global ones.

Parallel to the evolution of radio-mobile systems, we assist to the massive diffusion of Internet network and contents, thus allowing many users on the earth to be interconnected and to exchange any kind of information, data, images and so on.

Hence, there is a quick convergence of mobile communications and Internet, i.e., *mobile computing* (see Fig. 1).

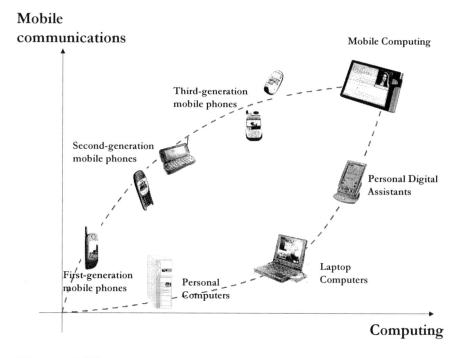

Fig. 1: Mobile communications and computing convergence towards mobile Internet access.

The first cellular systems became operational at the beginning of 1980 (first-generation, 1G). They employed analog techniques and rapidly diffused with each country having its own system. A first evolution was achieved 10 years later by the adoption of digital standards (second-generation, 2G). Presently, we are assisting to the deployment of third-generation mobile cellular systems (3G) that under umbrella recommendations collect at least three different standards. They are intended to provide the users with high bit-rate transmissions so as to allow a fast access to the Internet and, in general, multimedia transmissions on the move [i],[ii].

In some European countries and in Japan the widespread diffusion of mobile communications has reached the point to surpass the number of wired phones. This is an important achievement that significantly highlights the diffusion of mobile communication systems.

The unique capabilities of new cellular systems are expected to provide users with integrated multimedia applications. Small, powerful, application-enabled devices will bring mobility needs together with the desire for data and information. Networks will be based on the IP protocol [iii], including the support of *Quality of Service* (QoS) for differentiated traffic classes.

The air interface still represents the system bottleneck, by limiting the available user bit-rate due to both spectrum availability and radio propagation impairments.

At present, some mobile terminals have integrated a *Java Virtual Machine*, an important step towards the mobile computing and the support of typical Internet applications. In fact, the Java language permits the development of platform-independent applications. Another powerful tool for the realization of new applications and services is represented by the *eXtensible Markup Language* (XML) and related technologies. In fact, XML can be used to design Web pages that can be adapted to different Internet access devices and technologies (e.g., mobile terminals with small displays, *Personal Digital Assistants*, common personal computes, etc.) by using the characteristics of the *HyperText Transfer Protocol* (HTTP). In fact, an Internet server can be equipped with an adaptation engine that recognizes the access technology according to suitable fields in the HTTP packet header; hence, different translation rules can be used to adapt the XML contents [iv].

However, the expected diffusion of new applications and multimedia services can be only reached trough a novel system design that takes into account all the communication aspects from the application layer to the physical one, according to the OSI standard reference model. This approach is particularly effective for the air interface. In fact, a user application cannot be designed without accounting for the limited bandwidth, error resilience and reduced display sizes on mobile terminals. In addition to this, the performance of the transport layer protocol (TCP) must be evaluated in the presence of air interface resource constraints and the related traffic must be suitably managed to avoid that transmission delays or channel impairments negatively affect the TCP throughput. Moreover, the network layer must account for user mobility and the consequent re-routing of information when a user changes its cell. The frequency of handoff procedures among adjacent cells will be exacerbated in future 3G micro-cellular systems. Hence, the handoff process needs to be particularly optimized to avoid the loss of information during handoffs. Finally, the medium access control layer must be able to integrate the support of different traffic classes, guaranteeing *ad hoc* QoS levels, fairness among users and high utilization of radio resources.

All these aspects call for solutions suitably developed for the air interface [v]. Therefore, the focus of this book is on the optimization of the protocols at different layers in order to achieve simultaneously the maximum utilization of radio resources and the maximum satisfaction of users, two aspects typically in contrast.

This book will cover different wireless communication scenarios and, in particular: 2.5G and 3G mobile communication systems (i.e., GPRS, UTRA-FDD and UTRA-TDD); 4G broadband wireless access systems (e.g., HIPERLAN/2); mobile satellite systems. A complete review of such systems is carried out in PART I. Then, PART II will first focus on both the performance evaluation of different resource management techniques for the above mentioned air interfaces and, then, will address the protocols at network and transport layers to allow the mobile access to the Internet (i.e., TCP/IP and WAP). Hence, we will consider the impact on the throughput of cellular systems due to both the user mobility and the transmission of data packets on error-prone channels.

References

[i]   M. Zeng, A. Annamalai, V. K. Bhargava, "Recent Advances in Cellular Wireless Communications", *IEEE Comm. Mag.*, pp. 128-138, September 1998.

[ii]   Ojanpera and R. Prasad. *Wideband CDMA for Third Generation Mobile Communications*. Artech House, October 1998.

[iii]   T. Robles, A. Kadelka, H, Velayos, A. Lappetelainen, A. Kassler, H. Li, D. Mandato, J. Ojala, B. Wegmann, "QoS Support for an All-IP System Beyond 3G", *IEEE Comm. Mag.*, pp. 64-72, August 2001.

[iv]   Network Working Group, "Hypertext Transfer Protocol - HTTP/1.1", (Web page) URL: http://ww.ieft.org/rcf/rcf2616.txt, June 1999.

[v]   M. N. Moustafa, I. Habib, M. Naghshineh, M. Guizani, "QoS-Enabled Broadband Mobile Access to Wireline Networks", *IEEE Comm. Mag.*, Vol. 40, No. 4, pp. 50-56, April 2002.

Chapter 1: Multiple access techniques for wireless systems

In a wireless communication system, radio resources must be provided in each cell to assure the interchange of data between the mobile terminal and the base station. Uplink is from the mobile users to the base station and downlink is from the base station to the mobile users. Each transmitting terminal employs different resources of the cell. A *multiple access* scheme is a method used to distinguish among different simultaneous transmissions in a cell. A *radio resource* can be a different time interval, a frequency interval or a code with a suitable power level. All these characteristics (i.e., time, frequency, code and power) univocally contribute to identify a radio resource [1]. If the different transmissions are differentiated only for the frequency band, we have the *Frequency Division Multiple Access* (FDMA). Whereas, if transmissions are distinguished on the basis of time, we consider the *Time Division Multiple Access* (TDMA). Finally, if a different code is adopted to separate simultaneous transmissions, we have the *Code Division Multiple Access* (CDMA). However, resources can be also differentiated by more than one of the above aspects. Hence, hybrid multiple access schemes are possible (e.g., FDMA/TDMA).

In a cellular system, radio resources can be *re-used between sufficiently far cells*, provided that the mutual interference level is at an acceptable level. This technique is adopted by FDMA and TDMA air interface, where the reuse is basically of carriers. In the CDMA case, the number of available codes is so high that the code reuse among cells (if adopted) does not increase the interference.

In uplink, a suitable *Medium Access Control* (MAC) protocol is used to regulate the access of different terminals to the resources of a cell that are provided by a multiple access scheme [2]. Whereas, in downlink the base station has to transmit to the different users by means of a suitable multiplexing scheme. In the case of packet-switched traffics, there is also a packet scheduling function that has to be implemented in the base station.

The classical multiple access techniques are described below [1].

1.1   *Frequency Division Multiple Access* (FDMA)

The frequency band available to the system is divided into different portions, each of them used for a given channel (Fig. 1); the different channels are distributed among cells (according to a reuse pattern). Adjacent bands have guard spaces in order to avoid inter-channel interference. First-generation terrestrial cellular systems (such as *Advanced Mobile Phone System*, AMPS, that started operations in USA on 1979) were based on analog transmissions with frequency modulation and FDMA [3]. With the evolution towards digital communications, also TDMA and CDMA access schemes can be implemented.

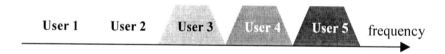

Fig. 1: FDMA technique.

One disadvantage of FDMA is the lack of flexibility for the support of variable bit-rate transmissions, an essential prerequisite for future mobile multimedia communication systems.

1.2   *Time Division Multiple Access* (TDMA)

In this scheme, each user has assigned the total bandwidth of a carrier for transmission, but only for a short time interval (*slot*) that is periodically repeated according to a time-organization called *frame*.

Transmission is organized into frames, each of them containing a given number of slot intervals, N_s, to transmit *packets* of bits (Fig. 2).

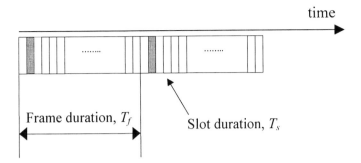

Fig. 2: TDMA frame.

For instance, let us refer to the transmission of speech through a digital communication system. The voice source signal (analogue signal) is sampled with a suitable rate. Each obtained value is then quantized with a suitable number of bits. Then, a source coding scheme can be adopted to reduce the transmission bit-rate. Finally, dynamic compression and predictive schemes are adopted (accordingly, it is possible to achieve a low bit-rate voice transmission up to 2.4 kbit/s, for some satellite systems). Thus, information bits are grouped in packets. A voice source typically require one packet to be transmitted a in a slot per frame (see the darkest slots in Fig. 2).

The US digital standard for cellular communications named IS-54 is based on TDMA and tripled the capacity (= number of simultaneous users supported per cell) with respect to the AMPS system, at a parity of total bandwidth [3]. The pan-European standard of second-generation cellular systems, GSM (*Global System for Mobile Communications*), is based on TDMA. More exactly, GSM adopts a hybrid scheme of the FDMA/TDMA type: the available bandwidth is divided among different 200 kHz sub-bands, each of them occupied by a carrier accessed with a TDMA scheme.

The main disadvantage of TDMA air interfaces is the high peak transmit power that is required to send packets in the assigned slots. Moreover, a fine synchronization must be achieved at the beginning of each transmission for the alignment with the time-frame structure. Finally, a rigid resource allocation is supported by TDMA: according to the above example of the voice traffic, one slot is assigned to a voice source also during silent periods among talkspurts.

1.3   Resource reuse with TDMA and FDMA

Cellular systems with TDMA or FDMA techniques are based on the resource reuse concept. Indeed, due to the limited number of radio resources, it is necessary to reuse the same resource among sufficiently distant cells so that the inter-cell interference is negligible. The *reuse distance D*, is the distance between two cells that may simultaneously use the same channel (see Fig. 3). Assuming a hexagonal regular cellular layout for a given D value, it is possible to divide the total number of resources into K groups, distributed among the different cells as in a mosaic. Possible values of K are: 1, 3, 4, 7, 9, …

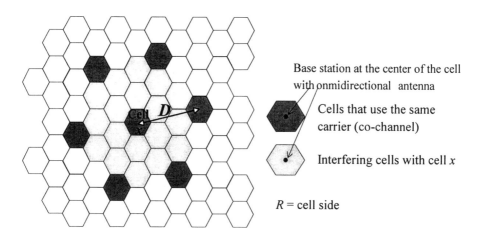

Fig. 3: Reuse of resources for a given reuse distance D.

It is possible to prove the following relationship among D, R (the cell radius) and factor K [4]:

$$D = R \sqrt{3\,K^2} \qquad (1)$$

The ratio between the power received at the base station from the desired user of its cell, C, and the power received from co-channel users in cells at distance D, I, can be expressed as follows:

$$\frac{C}{I} = \frac{1}{6}\left(\frac{D}{R}\right)^{\gamma}$$

(2)

where γ is the path loss exponent (varying from 2 to 4, depending on the cellular environment; $\gamma = 2$ is typical of free space propagation).

For $K = 7$, we obtain the reuse mosaic shown in Fig. 4 that corresponds to C/I = 18 dB for $\gamma = 4$.

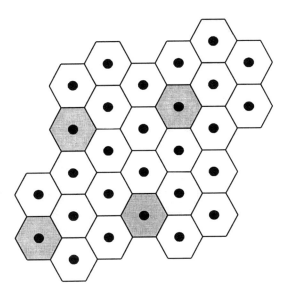

Fig. 4: Reuse distance pattern for $K = 7$.

For the sake of completeness it is important to stress the fact that in practical cases the cellular coverage is not hexagonal regular, but depends on streets, building heights, obstacles, and so on. A typical example can be the GSM 900 MHz cellular coverage shown in Fig. 5, where different colors are related to different cells irradiated by three-sectored sites.

Fig. 5: GSM 900 MHz radio cell coverage (city of Siena, Italy).

Let us refer to the classical circuit-switched voice service. On the basis of the reuse pattern K, if we have S system channels (i.e., frequency bands with FDMA or slots with TDMA), we may assign $Q = S/K$ resources per cell (*fixed channel allocation*). Hence, at most Q simultaneous circuit-switched phone calls can be managed per cell. A call generated in a cell where all its Q resources are busy is blocked and cleared. If we assume that calls arrive in a cell according to a Poisson process with mean rate λ and that the channel holding time in a cell, X, is *generally* distributed with mean value $E[X]$, the blocking probability P_b experienced by a call is given by the well-known ERLANG-B formula, according to an $M/G/Q/Q$ model (M stands for Poisson arrivals; G means a general call duration time distribution; Q is the number of requests in service = number of requests that can be hosted by the system) [5]:

$$P_b(\rho, Q) = \frac{\rho^Q}{Q! \sum_{n=0}^{Q} \frac{\rho^n}{n!}} \qquad (3)$$

where $\rho = \lambda E[X]$ Erlang.

The maximum cell capacity can be determined as the maximum load in Erlang, $\rho_{max}(Q)$, per cell that can be managed guaranteeing (for instance) $P_b \leq 1\%$. If each user contributes an elementary load of ρ_{user} Erlang, we may determine the maximum capacity of users per cell, $M_{max}(Q)$, as:

$$M_{max}(Q) = \frac{\rho_{max}(Q)}{\rho_{user}}: \text{ with } \rho_{max}(Q) \text{ the maximum } \rho \text{ value so that } P_b(\rho, Q) \leq 1\% \qquad (4)$$

Fig. 6 shows the behavior of $M_{max}(Q)$ as a function of Q assuming that each user contributes a load of 40 mErlang.

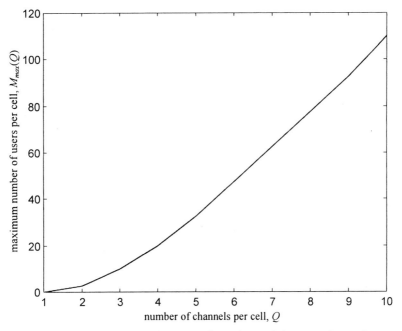

Fig. 6: User capacity behavior as a function of the number of resources per cell for target blocking probability of 1%.

1.4   *Code Division Multiple Access* (CDMA)

The concept at the basis of CDMA spreading the transmitted signal over a much wider band ("*Spread Spectrum*", SS). Such techniques were developed as jamming countermeasures for military applications in the years 1950. Accordingly, the signal is spread over a band *PG* times greater than the original one, by means of a suitable modulation based on a *PseudoNoise* (PN) code[1] [6]-[9].

PG is the so-called *Processing Gain*. The higher PG, the higher the spreading bandwidth and the greater the system capacity, as explained later in this Section. Each user has its own code for uplink transmissions. In downlink, each base station has its code, but, in addition to this, suitable codes must be used to distinguish the different simultaneous transmissions to the users in the cell.

Even if a concentrated interfering signal is present in a portion of the bandwidth of the spread signal, the receiver de-spreads the useful signal and spreads on a wide band the interfering one, so that it becomes more similar to background noise.

The receiver must use a synchronous code sequence with that of the received signal in order to correctly de-spread the desired signal.

There are two different techniques to obtain spread spectrum transmissions:

-   *Direct Sequence* (DS), where the user signal is multiplied by the PN code with bits (named *chips*) whose length is basically PG times smaller that that of the original bits. This spreading scheme is well suited for *Phase Shift Keying* (PSK) and *Quadrature Phase Shift Keying* (QPSK) modulations (see Fig. 7).

-   *Frequency Hopping* (FH), where the PN code is used to change the frequency of the transmitted symbols (see Fig. 8). We have fast

---

[1] PN codes are cyclic codes that well approximate the random generation of 0 and 1 bits (e.g., Gold codes). These codes must have a high peak for the auto-correlation (synchronization purposes) and very low cross-correlation values (for the orthogonality of different users).

hopping if frequency is changed at each new symbol; whereas, a slow hopping pattern is obtained if frequency varies after a given number of symbols. The *Frequency Shift Keying* (FSK) modulation is well suited for the FH scheme.

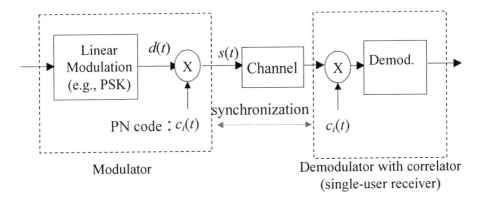

Fig. 7: Spreading and de-spreading processes for the *i*-th DS-CDMA user.

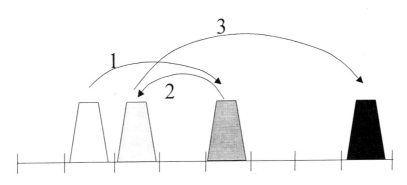

Fig. 8: Spreading process with FH-CDMA.

A significant advantage of spread spectrum techniques for mobile communications is that they allow transmissions that are particularly resistant to multipath fading (produced by reflections and diffraction of the signal due to the presence of obstacles along the signal path). In fact, spread transmissions by their nature mitigate the frequency selective affects due to multipath fading [10].

The DS-CDMA technology is preferred to the FH-CDMA one, since it is expensive to realize frequency synthesizers able to switch rapidly the transmission frequency.

With DS-CDMA, a useful signal in a cell can be perfectly separated from other DS-CDMA signals with different codes (interfering signals) in case of synchronous transmissions with orthogonal codes (null cross-correlation). If such synchronism is lost, partial cross-correlations among different codes loose the orthogonality, so that *Multiple Access Interference* (MAI) is experienced: the de-spreading process is unable to conceal completely the interference coming from simultaneous users in the cell. This is the most common case in DS-CDMA cellular systems. Referring to uplink, MAI contributions come from simultaneous transmissions in the same cell of the desired user and from adjacent cells. Note that synchronous transmissions can be naturally achieved for downlink transmissions in a cell. However, multipath phenomena may still introduce some intra-cell MAI.

Any technique able to reduce MAI increases capacity with DS-CDMA. In particular, we may consider:

- Squelching of transmissions during inactivity phases;

- Multi-sector cells with directional antennas at the base station;

- Multi-user receivers that reduce MAI coming from the users in the same cell (intra-cell interference).

With CDMA transmissions, it is possible to use a special receiver, named RAKE, that combines the signal contributions coming from different paths (micro-diversity). This receiver is particularly useful in the multipath environment of mobile communications in order to improve the bit error rate performance [10].

CDMA well supports powerful coding schemes that partly contribute to the spreading process. Accordingly, CDMA permits to achieve a greater robustness and a higher capacity than other multiple access schemes (i.e., TDMA and FDMA). Hence, CDMA has been selected for future-generation mobile communication systems.

CDMA needs that a power control scheme be adopted in order to avoid that a user closer to the base station be received with an overwhelming power with respect to users at cell borders (*near-far problem*) [8]. Hence, the signals of all the users must be received with the same power level (both for uplink and downlink), unless complex multi-user receivers are adopted.

In general, multipath, shadowing and path loss phenomena call for a continuous regulation of the transmitted power. Channel propagation variations are related to the Doppler frequency. In *Open-Loop Power Control* (OLPC) schemes, the transmitter adapts the emission power depending on the power level of the received signal. In *Closed-Loop Power Control* (CLPC) schemes the receiver notifies the received power level to the sender that, thus, may vary the transmitted power to guarantee an adequate received level. OLPC and CLCP can be also adopted to maintain a given received *Signal-to-Noise and Interference Ratio* (i.e., SNIR-based power control schemes).

For example, OLPC can be useful for quiescent mobile terminals that continuously measure a *beacon signal* (*i.e., the pilot signal*) broadcast by the base station in order to regulate the power they use for the first access. Such technique is adopted for random access transmissions (i.e., first attempt after a silence phase). Whereas, CLPC can be adopted when a continuous communication is established between a base station and a mobile terminal.

Third-generation cellular systems adopt a SNIR-based CLPC scheme based on two different cooperating loops: the *inner* loop and the *outer* one. In particular, the outer loop continuously measures the signal quality and defines the SNIR level value to be achieved (i.e., SNIR target) to guarantee a given error rate performance. The inner loop continuously updates the transmission power level so as to maintain the defined SNIR target value (one power control update measure is sent every 0.665 ms in the WCDMA 3G system and every 1.25 ms in the cdma2000 3G system).

1.4.1 DS-CDMA spreading process

We consider a PSK modulation where antipodal rectangular bits (i.e., +1 and -1) of duration T_b are transmitted by multiplying them with a carrier oscillation (see Fig. 7). Before sending the signal, the bits are modulated by a chip code sequence, where each chip has duration T_c $\ll T_b$. This process is detailed in Fig. 9, where we assume that the PN code is a periodic sequence with period corresponding to the bit duration and that bits and chips have rectangular shapes (i.e., roll-off factor equal to 0, for an ideal case).

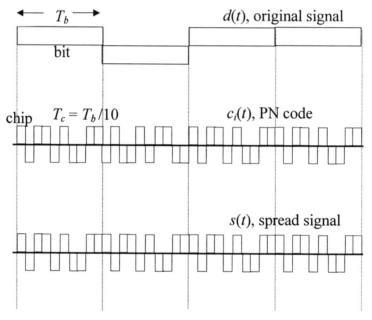

Fig. 9: Base-band spreading process for DS-CDMA. Signals $d(t)$, $c(t)$ and $s(t)$ are related to Fig. 7. In this example, PG = 10.

The processing gain PG is obtained as (see Fig. 9):

$$PG = \frac{T_b}{T_c} \qquad (5)$$

Typically, PG $\gg$ 1.

The base-band spreading process and the related frequency spectrum is further described in Fig. 10.

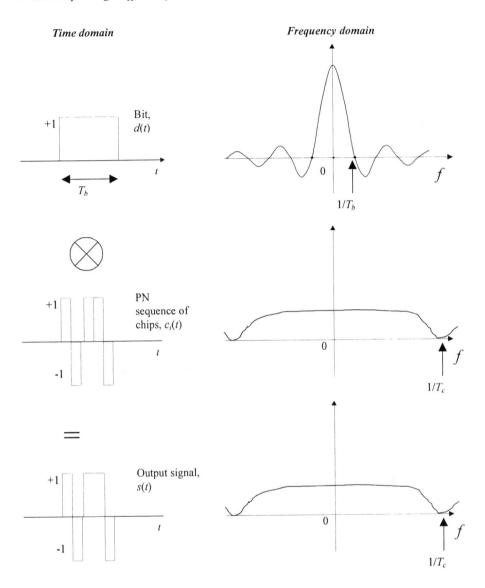

Fig. 10: Time and frequency representation of the signals related to the DS spreading process (base-band signals).

1.4.2 Basic considerations on the capacity of DS-CDMA systems

We refer here to uplink and we assume that simultaneous transmissions in the same cell and in adjacent cells contribute a MAI level that can be modeled as *Additive White Gaussian Noise* (AWGN channel assumption). Let us consider having M equal mobile terminals perfectly power-controlled in each cell of the DS-CDMA cellular systems (single traffic case). We can write the following SNIR expression, assuming a single-user correlation receiver and non-synchronous received signals at the base station:

$$SNIR = \frac{P}{I+N} = \frac{P}{I_{intra-cell} + I_{inter-cell} + \eta_0 W} = \frac{P}{(M-1)P + \varepsilon MP + \eta_0 W} \quad (6)$$

where:

- P is the received power from a power-controlled user
- ε is (approximately) the inter-cell to intra-cell interference ratio (ε depends on both channel characteristics and power control scheme; a typical ε value is around 0.6, but even higher values are possible [11],[12])
- W is the spreading bandwidth
- η_0 is the single-sided background noise power spectral density.

We may notice that $P = E_b R_b$ (energy per bit multiplied by the bit-rate) and we may consider $I + N$ as the power of a white Gaussian noise over the bandwidth W with a given density I_0. Through some algebraic manipulation, from (6) we obtain:

$$\frac{E_b}{I_0} \approx \frac{\frac{W}{R_b}}{(1+\varepsilon)M + \frac{\eta_0 W}{P}} \cdot \quad (7)$$

Note that $R_b = 1/T_b$ and $W = (1 + \alpha)/T_c$, where α il the roll-off factor of the impulse adopted to transmit a chip. Hence, the spreading term W/R_b in (7) can be expressed as follows:

$$\frac{W}{R_b} = (1+\alpha)PG \cdot \quad (8)$$

By substituting (8) in (7), we have:

$$\frac{E_b}{I_0} \approx \frac{(1+\alpha)PG}{(1+\varepsilon)M + \frac{\eta_0 W}{P}} \qquad . \tag{9}$$

We impose $E_b/I_0 \geq E_b/I_0 \,|\, _{target}$ as *Quality of Service* (QoS) requirement: $E_b/I_0 \,|\, _{target}$ is set to guarantee a given bit error rate value. We assume that the background noise level is negligible with respect to the useful signal: $N = \eta_0 W \ll P$. Moreover, we may consider that each user transmits according to an activity factor ψ, therefore, in previous formulas, we may substitute ψM to M. Thus, (9) may be solved with respect to M in order to find the maximum capacity of simultaneous users per cell:

$$\frac{(1+\alpha)PG}{(1+\varepsilon)\psi M} \geq \frac{E_b}{I_0}\bigg|_{target} \quad \Rightarrow \quad M \leq \frac{(1+\alpha)PG}{(1+\varepsilon)\psi \dfrac{E_b}{I_0}\bigg|_{target}} \qquad . \tag{10}$$

The user capacity limit in DS-CDMA systems $\dfrac{(1+\alpha)PG}{(1+\varepsilon)\psi \dfrac{E_b}{I_0}\bigg|_{target}}$ increases with the processing gain and decreases

with both $E_b/I_0 \,|\, _{target}$ and the inter-cell interference factor ε.

The base station must be provided with an adequate number of receivers in order to support the capacity expressed by (10). If a limited number of receivers is present, call blocking events may occur.

CDMA systems adopt the soft-handoff scheme, so that while a user change a cell, the signals received from (sent to) two cells are combined in downlink (uplink) to improve the transmission quality by means of some degree of diversity. Such technique improves the communication quality during a handoff, but it requires both a greater number of receivers (to avoid blocking events) and adequate power control schemes. We can roughly state that if 20% of cell area contains users in soft-handoff, given M simultaneous users per cell, by reciprocity, the number of receivers at the base station must be $M + M/5$ to support all the transmissions (including the duplicated ones due to soft-handoffs).

Note that if a multi-user receiver is used at the base station, we may neglect the intra-cell interference contribution in (6), so that the cell capacity limit becomes:

$$M \leq \frac{(1+\alpha)PG}{\varepsilon \psi \left. \frac{E_b}{I_0} \right|_{target}} \cdot \qquad (11)$$

This is also the typical condition for downlink transmissions, where we may assume that orthogonality among users is preserved (with a good approximation, especially in micro-cellular systems[2]). Downlink capacity is not interference limited, but rather power limited, due to the fact that the base station has to divide its available power among all the $M\psi$ simultaneously transmitting users. Hence, high power levels are needed in the presence of far users, thus reducing the capacity of simultaneous users if equal power levels have to be guaranteed for all the transmissions of a cell to avoid near-far problems with single-user receivers.

---

[2] In macro-cellular systems, heavy multipath phenomena in urban areas may introduce intra-cell interference in downlink.

Chapter 2: The Global System for Mobile Communications

The GSM system originally defined in the 1980s was intended as a Pan-European standard. Today GSM has expanded into many parts of the World: GSM is widely adopted not only in Europe, but also in Australia, Hong Kong, Singapore, South Africa and the UAE [13].

To provide additional capacity and to enable higher subscriber densities, two other systems were added later: GSM1800 (also known as DCS1800) and GSM1900 (also named PCS 1900). Compared to GSM900, GSM1800 and GSM1900 differ primarily in the air interface. Besides adopting other frequency bands, they use a microcellular structure (i.e., a smaller coverage radius for each cell) thus achieving a closer reuse of resources and, then, a higher capacity. A very accurate history on the GSM standard can be found in [14]. A complete survey of other 2G systems is shown in [3].

2.1 Introduction to GSM

The GSM system is characterized by the building blocks that are detailed below together with their interrelations, according to a typical architecture that distinguishes between the base station sub-system, the network sub-system and the mobile station (see Fig. 11) [14].

2.1.1 Base station sub-system

Base Transceiver Station (BTS) is the base station with an antenna to cover a cell (or more sectors).

The *Base Station Controller* (BSC): a group of BTSs is connected to a particular BSC, which manages their radio resources. The primary function of the BSC is call maintenance. *Mobile Stations* (MSs) send reports of their received signal strengths to the BSC every 480 ms. With this information the BSC decides to initiate handovers to other cells, to change the BTS transmission power, etc.

2.1.2   Network sub-system

The *Mobile Switching service Center* (MSC) acts like a standard exchange of the *Integrated Services Digital Network* (ISDN) and additionally provides all the functionality needed to support user mobility. The main functions are: registration, authentication, location updating, handovers and call routing to a roaming subscriber. If the MSC has also interconnections towards other networks, it is called *Gateway MSC* (GMSC).

The *Home Location Register* (HLR) is a database used for the management of mobile subscribers. It stores the *International Mobile Subscriber Identity* (IMSI), *Mobile Station ISDN Number* (MSISDN) and the address of the current *Visitor Location Register* (VLR). HLR contains the information to route the call to the VLR where the destination MS is currently registered. The HLR also maintains the description of the services associated with each mobile user (*profile*).

For each MS currently located in the geographical area controlled by the VLR, the VLR contains the current MS location and selected administrative information from the HLR, necessary for call control and the provision of the subscribed services. A VLR is connected to one MSC and is normally integrated into the MSC hardware.

The *Authentication Center* (AuC) is a protected database that holds a copy of each subscriber SIM card secret key that is used for authentication and encryption over the radio channel. AuC is normally located close to each HLR within a GSM network.

The signaling between functional registers in the network sub-system uses *Signaling System 7* (SS7).

The *Short Message Service Center* (SMSC) enables subscribers to send and receive SMS (*Short Message Service*) messages in the cellular network. The SMSC temporarily stores SMS than cannot be delivered due to an unreachable user (store-and-forward service). The SMSC is linked to an MSC, through which it sends and receives SMS. The SMSC obtain SMS routing information from the HLR.

The *Equipment Identity Register* (EIR) is a database that contains a list of all valid mobile station equipment within the network, where each mobile station is identified by its *International Mobile Equipment Identity* (IMEI). EIR is formed by three databases:

- *White list* for all known, good handsets

- *Black list* for bad or stolen handsets

- *Grey list* for handsets that are uncertain.

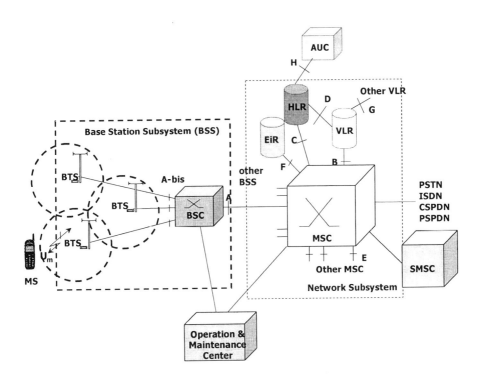

Fig. 11: Classical GSM system architecture.

The GSM core network is based on *Mobile Application Part* (MAP) protocols [14].

The *International Telecommunication Union* (ITU), which manages international allocation of radio spectrum (among many other functions), has allocated the following bands:

GSM900
Uplink: 890 - 915 MHz (= mobile station to base station)
Downlink: 935 - 960 MHz (= base station to mobile station)

GSM1800
Uplink: 1710 - 1785 MHz
Downlink: 1805 - 1880 MHz

GSM1900
Uplink: 1850 - 1910 MHz
Downlink: 1930 - 1990 MHz.

In conventional GSM, a radio channel is permanently allocated for a particular user during the entire call period (whether or not data is transmitted). The following services are supported:

- Voice service

- SMS

- Fax service

- Data service

- Transparent data (no error correction and constant delay)

- Non-transparent data (error correction and variable delays)

- Access to modems in telephony networks; i.e., access to the Internet requires a circuit-switched call to an *Internet Service Provider* (ISP).

2.2   GSM standard evolution

With GSM-phase 1 (frozen on 1990), there is a fixed capacity assigned for all the duration of a call according to circuit-switching. Users pay on the basis of the connection time and not according to the traffic volume transmitted during the call. Connection setup takes time. Data transmissions are slow (at 9.6 kbit/s). Therefore, GSM-phase 1 was mainly conceived for speech transmissions.

A first evolution of the GSM standard was GSM-phase 2 (completed on 1995), with the introduction of packet-oriented data transmission services such as SMS and *Unstructured Supplementary Service Data* (USSD):

- An SMS contains at most 160 characters (140 octets) sent to/from an MS via a signaling channel that depends on the status of the MS; SMS is a store-and-forward service provided by an SMSC: messages are kept within the SMSC until delivered to MS. Paging of the MS is needed to send each SMS message. The bit-rate for the transmission of an SMS is about 800 bit/s.

- Each USSD message contains at most 140 octets, conveyed on the air interface as SMS with bit-rate approximately of 800 bit/s. USSD is a transaction-oriented service with multiple mobile-originated or mobile-terminated messages during one session. USSD traffic is always exchanged with the HLR. The GSM network can route USSD data to an external server.

SMS can be used for applications like news, weather, stock exchange or road traffic information. Whereas, USSD are well suited for transactional applications.

The great success of SMS traffic highlights the need for packet data transmissions on the move. At present, 1 billion SMS messages are sent every day. It is expected that there will be 1.1 billions Internet users by 2004 and at least half of them will use wireless terminals to access the Web. A first answer to these needs is represented by the *General Packet Radio Service* (GPRS) [15]-[20] that provides packet switched services in an evolved GSM network (GSM-phase 2+) up to (theoretically, with no coding protection) about 170 kbit/s by assigning 8 slots of the same frame to a given user. However, present technological implementations allow up to 4 slots to be destined to the same mobile terminal, thus achieving the potential maximum bit-rate per user of about 85 kbit/s.

In order to reuse frequencies, GSM and GPRS adopt a combination of FDMA and TDMA (i.e., reuse of carriers, each with TDMA resource sharing scheme).

2.3   GPRS network architecture

GPRS introduces new network nodes in the existing GSM system architecture. The most important ones are the *Serving GPRS Support Node* (SGSN) and the *Gateway GPRS Support Node* (GGSN). However, the GPRS architecture also introduces the following elements:

- *Point-to-Multipoint Service Center* (PTM-SC),
- *Border Gateway* (BG),
- *Charging Gateway* (CG),
- *Legal Interception Gateway* (LIG).

A complete GSM-GPRS architecture is shown in Fig. 12.

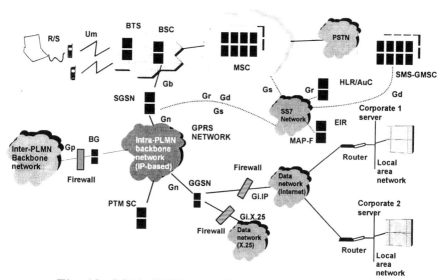

Fig. 12: GSM-GPRS complete network architecture.

Whereas, a simplified GPRS architecture is given in Fig. 13, where the following interfaces are shown (for the description of some related protocols, see the following Section 2.8):

- Gb LLC (user data) and BSSGP (signaling) over *Frame Relay* (FR)
- Gc *Mobile Application Protocol* (MAP); for location information retrieval
- Gd for short messaging over GPRS

- Gf MAP for checking the mobile equipment identities
- Gn *GPRS Tunneling Protocol* (GTP) for intra-PLMN traffic
- Gp GTP (over IP) for inter-PLMN traffic
- Gr MAP to access subscriber information
- Gs BSSAP plus protocol for normal location updates and paging via MSC/VLR
- Gi IP (or X.25) protocol interface to external data networks.

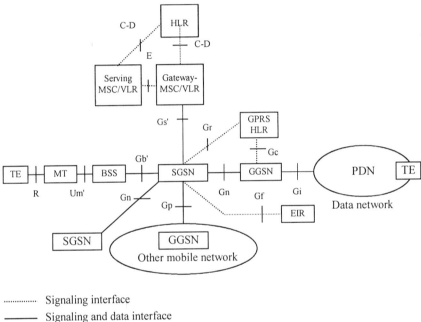

Fig. 13: GPRS network interfaces (only packet-switched domain).

Table 1 below describes some technological characteristics of the interfaces in the GPRS network.

The SGSN is an important component of the GPRS network and operates at the same hierarchical level of the MSC in the classical circuit-switched GSM domain. In particular, the SGSN manages user mobility, authentication and security. The SGSN is connected to BSCs and represents the service access point to the GPRS network for GPRS-enabled MSs. The connection between SGSN and BSC is based on

Frame Relay (FR), as detailed in Table 1. The SGSN handles the conversion from the IP protocol used in the GPRS backbone network to the SNDCP and LLC protocols used between the SGSN and the MS. These layers handle compression and ciphering. When the MS wants to send (or receive) data to (from) external networks, data are relayed between the SGSN and the relevant GGSN (and vice versa).

Interface	Between Entities	Technology	Physical
G_b	SGSN and BSC	FR	T1
G_d	SGSN and SMS	SS7	T1
G_i	GGSN and PDN	IP	fiber
G_n/G_p	GSNs within/inter- PLMN	IP	Ethernet, T1, fiber
G_r/G_c	SGSN/GGSN and HLR	SS7	T1
G_s	SGSN and MSC/VLR	SS7	T1

Table 1: Networking technologies for GPRS interfaces (T1 = PCM scheme at 1.5 Mbit/s [21]).

The GGSN provides interworking capability with external packet data networks, like X.25 and IP networks. The GGSN contains routing information for all the attached GPRS users. Therefore, from external network point of view, the GGSN is similar to a router. When the GGSN receives data addressed to a specific user, it checks if the related address is active. If this is the case, the GGSN forwards the data to the SGSN serving the MS (the GGSN is connected to SGSN by private IP-based network); if the address is inactive, data are discarded. Each network has a pool of available Internet addresses that are dynamically assigned to the users by GGSN. A dynamic address is allocated to an MS only for the duration of a connection. The GGSN tracks the MS with a location precision at the SGSN level.

Both SGSN and GGSN send *Call Detail Records* (CDRs), corresponding to traffics generated in opened sessions to a *Charging*

Gateway (CG) that sends these data to an external *customer care and billing system* for further processing.

The *Lawful Interception Gateway* (LIG) provides access to law enforcement authorities to analyze packet data traffic from the selected MS.

The *Point-To-Multipoint Service Center* (PTM-SC) is similar to the SMSC in the classical GSM network, with the exception that it uses the GPRS transport network.

Three different MS types are considered in the GPRS system:
- *Class A*: simultaneous GPRS and conventional GSM operations
- *Class B*: a terminal of this class can be attached to either GPRS or conventional GSM services (either circuit-switched calls or GPRS data transfer, but no simultaneous use of services)
- *Class C*: a terminal that can be only attached to GPRS.

2.4   GSM-GPRS air interface: details on physical layer

The whole GSM frequency band is split by using FDMA into 124 channels, each 200 kHz wide. Each channel is further divided using TDMA into 8 time slots (see Fig. 14). A time slot lasts 577 µs and contains 156.25 bits [14]. Hence, the entire frame length is 4.613 ms. The air interface bit-rate is 270.844 kbit/s. A physical channel is defined by the recurrence of a time slot in subsequent frames.

In the classical GSM, a group of 26 TDMA frames are combined to form a 26-frame multi-frame, which lasts for 120 ms. The multi-frame consists of 24 frames that are used for traffic, 1 frame that is used for the *Slow Associated Control Channel* (SACCH) and 1 unused frame.

Let us refer to classical GSM circuit-switched data traffic: the user can only transmit during one slot of the TDMA frame. Each burst period contains 114 bits of data; therefore, in each multi-frame a user can transmit a total of 2736 bits. Out of these bits, only 52.63% are user data. Therefore, a user can only transmit 1440 bits of data per multi-frame. Since a multi-frame is transmitted in 120 ms, the effective radio interface rate is 12 kbit/s. However, the net bit-rate that can be achieved is only 9.6 kbit/s due to adaptation functions at the MS - Base Station sub-system interface.

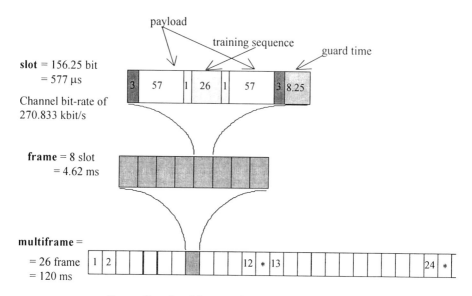

slot = 156.25 bit
     = 577 μs

Channel bit-rate of
270.833 kbit/s

frame = 8 slot
     = 4.62 ms

multiframe =

= 26 frame
= 120 ms

Frames from 1 to 24 are used to convey user information;
Frames with "*" carry signaling traffic or are unused.

Fig. 14: GSM TDMA organization.

In GPRS, a group of 52 TDMA frames are combined to form a 52-frame multi-frame [22]. Out of these 52 frames, 48 are used for data, 2 are used for the *Packet Timing Advance Control Channel* (PTCCH) and 2 frames are idle.

The basic radio packet in GPRS is the RLC/MAC (*Radio Link Control/Medium Access Control*) block: it uses a sequence of four time slots in four subsequent frames, as shown in Fig. 15. Different logical channels can share the same *Packet Data physical CHannel* (PDCH), i.e., a slot. The PDCH sharing is on a radio block basis.

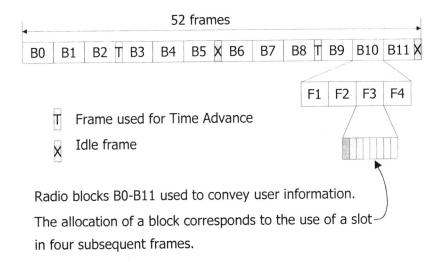

Fig. 15: Time organization of a slot (i.e., *Packet Data physical CHannel*, PDCH) for GPRS.

GPRS is a packet-switched system specifically tailored for data traffic: channels (i.e., slots) are only assigned to a user for the duration of its transmission. As a result, GPRS may allot all the eight time slots of a frame on a given carrier to a user, depending on the load of the system.

GPRS supports four different coding schemes that are tailored for different channel conditions (see Table 2) [23]:

- CS-1 uses half-rate convolutional coding; it is the most robust scheme, used when there are bad channel conditions. CS-1 supports a data rate of 9.05 kbit/s per time slot in the frame.

- CS-2 allows a data rate of 13.4 kbit/s, but it is a little less robust than CS-1.

- CS-3 is a less reliable coding scheme than CS-2, but it supports data rates of 15.6 kbit/s. CS-2 and CS-3 coding schemes are punctured versions of the CS-1 code.

- CS-4 scheme uses no error correction; therefore, it can only be used under good channel conditions. CS-4 supports data rates of 21.4 kbit/s per time slot.

The most typical coding schemes are CS-1 and CS-2. It is unlikely that CS-3 and CS-4 be implemented in the network due to the complexity for the management of CS-3 and CS-4 channels on the Abis interface (BTS-BSC); it is, in fact, preferable to allocate capacities in multiple of 13 kbit/s (CS-2) on Abis.

If a user has assigned all the eight time slots of a TDMA frame, the user can obtain a combined data rate of 171.2 kbit/s with the CS-4 scheme. To achieve this maximum rate the user must be able to fully utilize one GSM carrier (lightly loaded system) and must have good channel conditions.

Even if the theoretical maximum capacity for GPRS corresponds to the assignment of all the eight slots of a frame to a user, the practical maximum capacity corresponds to four slots. This is due to the fact that GSM phones are half duplex (they cannot simultaneously transmit and receive). Hence, in a frame, four slots are used to receive (downlink), one slot is for downlink-to-uplink switching operations, one slot is used for uplink transmissions and two slots may be used for measurements of neighboring cells. Consequently, the practical maximum GPRS capacity is about 53 kbit/s with CS-2.

Scheme	Code rate	Data rate per slot in the frame [kbit/s]
CS-1	1/2	9.05
CS-2	≈2/3	13.4
CS-3	≈3/4	15.6
CS-4	1	21.4

Table 2: GPRS coding schemes.

A packet format (with some overhead) corresponds to each layer of the GPRS protocol stack, as detailed in Fig. 16, where the packets at LLC (*Logical Link Control*) level are named *Service Data Units* (SDUs) or *LLC frames* and the packets at RLC/MAC level are named *Packet Data Units* (PDUs).

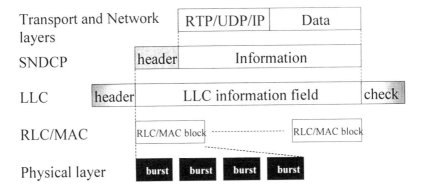

Fig. 16: Packet formats at different layers of the GPRS protocol stack
(see also the following sub-Section 2.8.2).

Let us consider a practical GPRS dimensioning example. We refer to a GSM cell with four transceivers (i.e., carriers). Hence, there are 32 resources (= slots), among them only 30 can be effectively considered to carry information traffic (the others are used for signaling traffic). Considering a cell planning made to guarantee a blocking of 1% for voice circuit-switched traffic, we have that the cell can support up to 20 Erlang of voice traffic according to the classical ERLANG-B formula. Hence, on average, there are 10 channels available for GPRS traffic in the cell. Assuming to use the CS-1 coding scheme, 9.05 kbit/s per channel are supported with a total capacity of 90.5 kbit/s per cell. If we consider that each user contributes a mean traffic of 5 kbit/s, a most 18 simultaneous GPRS users can be admitted in the cell. Finally, assuming that only 10-20% of the active GPRS users are simultaneously transmitting (each GPRS user produces a bursty traffic with activity factor 10-20% and, therefore, peak-to-mean traffic ratio of 5-10), the maximum bit-rate experienced by each user during its GPRS session ranges from 25 to 50 kbit/s.

2.5   EDGE and E-GPRS

EDGE (*Enhanced Data Rates for Global Evolution*) is a new radio interface technology that adopts a new modulation (i.e., 8-PSK) for the PDCH. EDGE is also part of the US IS-136 standard belonging to 3G systems, as detailed in the following Chapter 3. With this air interface, the system can provide a data rate up to 384 kbit/s and a spectrum efficiency of 0.5 bit/s/Hz/base. The term GERAN (*GSM/EDGE Radio Access Network*) generally indicates an access network of the EDGE or GSM-GPRS type. Enhanced GPRS (E-GPRS) is for GPRS adapted to EDGE. Enhanced SGSN and GGSN are called E-SGSN and E-GGSN. The coding schemes of EDGE are detailed in the following Table 3.

Scheme	Code rate	Modulation	Data rate kbit/s
MCS-8	1.0	8-PSK	59.2
MCS-7	0.76		44.8
MCS-6	0.49		29.6
MCS-5	0.37		22.4
MCS-4	1.0	GMSK	17.6
MCS-3	0.80		14.8
MCS-2	0.66		11.2
MCS-1	0.53		8.8

Table 3: E-GPRS coding schemes.

2.6   Radio resource management concepts

GPRS uses packet-switched resource allocation: physical resources (i.e., slots) are allocated to a user (uplink and/or downlink) only when data is to be sent/received. The allocation of physical resources (i.e., slots or PDCHs) is dynamic from one to eight time slots of the same carrier. Uplink and downlink slots are separately assigned.

A *Temporary Block Flow* (TBF) is a physical connection used to allow the transfer of a number of blocks, identified by *Temporary Flow Identifier* (TFI). The TFI is included in every transmitted block, so that multiplexing of blocks originated from different mobile stations is possible on the same PDCH. TFI is assigned by the BSS and is unique in each direction. A TBF is used in the RLC/MAC layer for the unidirectional transfer of LLC SDUs on packet data physical channels. A TBF is temporary, maintained only for the duration of the data transfer.

Physical channels available in the cells can be shared between circuit-switched and packet-switched traffic either in a fixed boundary or in a movable boundary way.

The simulation results shown in the following figures highlight the impact that the GPRS packet-switched traffic has on the blocking of the classical GSM circuit-switched phone calls (ERLANG-B model) referring to downlink. If a rigid separation of resources between GPRS slots and GSM ones is adopted (see Fig. 17), the call blocking probability significantly increases with the number of slots destined to GPRS traffic. This solution significantly worsens the performance of GSM circuit-switched traffic. Hence, we may consider another resource sharing approach where there is no rigid separation between GPRS slots and GSM ones: the allocation of physical channels to GPRS is dynamic ("capacity on demand"). In this case, we assume that voice traffic has a *preemptive resume priority* with respect to data traffic. The results shown in Fig. 18 highlight that the voice traffic performance is unaffected by the presence of data traffic. However, data traffic may experience higher delays. A good compromise between these two cases could be to assign statically a minimum number of resources for GPRS traffic, leaving the possibility to employ dynamically additional resources for GPRS, if they are not used by the highest priority voice traffic. Hence, when the GSM phone traffic is low, more resources can be temporarily allocated to GPRS traffic (if necessary).

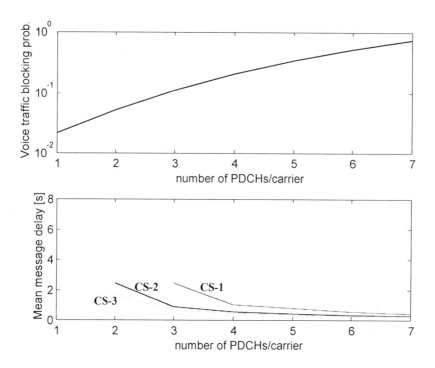

Fig. 17: GPRS slots and GSM slots as separated resources statically assigned (voice traffic load of 3 Erlang; data traffic with geometrically distributed message length with mean value of 10 kbit and Poisson arrivals with mean rate 2 msg/s).

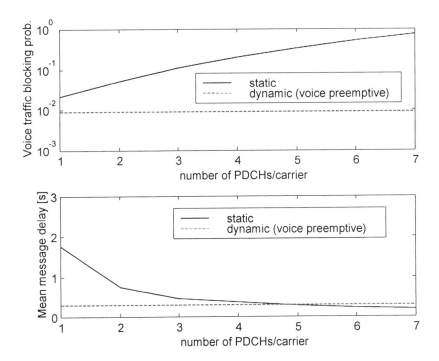

Fig. 18: Comparison between the static sharing of resources between GPRS slots and GSM slots and the dynamic sharing of resources with preemptive priority for voice (voice traffic load of 3 Erlang; data traffic with geometrically distributed message length with mean value of 10 kbit and Poisson arrivals with mean rate 0.5 msg/s; CS-3 coding scheme).

We consider the traffic balance for a given carrier. Let $\rho_{voice} = \lambda_{voice} T_{voice}$ denote the offered voice traffic intensity for a carrier in Erlang (being λ_{voice} the mean call arrival rate in call/s and T_{voice} the mean call holding time in seconds). Let $\rho_{data} = \lambda_{data} L / B_{CS}$ denote the data traffic intensity in pkts/slot (being λ_{data} the mean message arrival rate, L the mean message length in bits and B_{CS} the aggregated information bit-rate in bit/s of a carrier with a given coding scheme CS).

Considering a given carrier, we have $N = 8$ resources.

With the fixed resource sharing scheme between GSM and GPRS (as in Fig. 17), we have the following upper bound to the data traffic if $G \le N$ slots are destined to GSM traffic:

$$\rho_{data} < \frac{N-G}{N} = 1 - \frac{G}{N} \quad \left[\frac{\text{pkts}}{\text{slot}}\right] \Leftrightarrow N\rho_{data} < N-G \quad [\text{Erlang}] . \quad (12)$$

With the dynamic resource sharing scheme between GSM and GPRS (Fig. 18), we can write the following upper bound to the data traffic:

$$\rho_{voice}(1-P_b) + N\rho_{data} < N \quad [\text{Erlang}]$$

$$\Rightarrow \rho_{data} < 1 - \frac{\rho_{voice}(1-P_b)}{N} \quad \left[\frac{\text{pkts}}{\text{slot}}\right]. \quad (13)$$

where $P_b = P_b (\rho_{voice}, N)$ is the blocking probability given by the ERLANG-B formula shown in Chapter 1 of this Part.

2.7   QoS issues in the GPRS system

The QoS management enables the differentiation among provided services.

In GPRS Release 1998 five QoS attributes are defined [24]: precedence class, delay class, reliability class, mean throughput and peak throughput class. By the combination of these attributes many QoS profiles can be defined. Each attribute is negotiated by the MS and the GPRS network (see the following description of the *Packet Data Protocol*, PDP, context activation). If both parties accept the negotiated QoS profiles, the GPRS network will have to provide adequate resources to support these QoS profiles.

The RLC/MAC layer supports four radio priority levels and an additional level for signaling messages [22],[25]. See Table 4.

Precedence	Precedence interpretation
1 Highest priority	Service commitments shall be maintained ahead of precedence classes 2 and 3.
2 Normal priority	Service commitments shall be maintained ahead of precedence class 3.
3 Low priority	Service commitments shall be maintained after precedence classes 1 and 2.

Table 4: Precedence classes in GPRS.

Delay classes are identified as detailed in Table 5 [24] referring to end-to-end transfer time between two MSs, or between an MS and the Gi interface). We can note that there are four delay classes: the first three classes specify the supported values for both mean and the 95-th percentile of transfer delay, while the fourth class has unspecified values. This table specifies delay values for LLC SDU sizes of 128 and 1024 bytes.

	Delay (maximum values)			
	SDU size: 128 octets		SDU size: 1024 octets	
Delay Class	Mean Transfer Delay [s]	95th percentile Delay [s]	Mean Transfer Delay [s]	95th percentile Delay [s]
1.Predictive	< 0.5	< 1.5	< 2	< 7
2.Predictive	< 5	< 25	< 15	< 75
3.Predictive	< 50	< 250	< 75	< 375
4.Best Effort	Unspecified			

Table 5: GPRS end-to-end delay classes.

The reliability classes define the probabilities of loss, duplication, out of sequence and corrupted packets, as in Table 6.

The throughput parameter defines the peak octet rate per minute and the mean octet rate per second measured at Gi and R interfaces. The following classification applies.

Peak throughput class [24]: the peak throughput (Table 7) specifies the maximum rate at which data is expected to be transferred across the network for an individual session (= PDP context, as explained later).

There is no guarantee that this peak rate can be achieved or sustained for any time period, this depends upon the MS capability and available radio resources. The network may limit the subscriber to the negotiated peak data rate, even if additional transmission capacity is available.

Mean throughput class [24]: the mean throughput (Table 8) specifies the average rate at which data is expected to be transferred across the GPRS network during the remaining lifetime of an activated *Packet Data Protocol* (PDP) context. The network may limit the subscriber to the negotiated mean data rate (e.g., for flat-rate charging), even if additional transmission capacity is available. A best effort mean throughput class may be negotiated, meaning that throughput is made available to the MS on a per need and availability basis.

Reliability class	Lost SDU probab.	Duplicate SDU probab.	Out of Sequence SDU probab.	Corrupt SDU probab.	Example of application characteristics
1	10^{-9}	10^{-9}	10^{-9}	10^{-9}	Error sensitive, no error correction capability, limited error tolerance capability.
2	10^{-4}	10^{-5}	10^{-5}	10^{-6}	Error sensitive, limited error correction capability, good error tolerance capability.
3	10^{-2}	10^{-5}	10^{-5}	10^{-2}	Not error sensitive, error correction capability and/or very good error tolerance capability.

Table 6: Reliability classes for GPRS.

Peak throughput class	Peak throughput [octets/s]
1	Up to 1,000 (8 kbit/s)
2	Up to 2,000 (16 kbit/s)
3	Up to 4,000 (32 kbit/s)
4	Up to 8,000 (64 kbit/s)
5	Up to 16,000 (128 kbit/s)
6	Up to 32,000 (256 kbit/s)
7	Up to 64,000 (512 kbit/s)
8	Up to 128,000 (1,024 kbit/s)
9	Up to 256,000 (2,048 kbit/s)

Table 7: Peak throughput classes for GPRS (including also peak bit-rate classes for 3G systems).

Mean throughput class	Mean throughput [octets/hour]
1	Best effort
2	100 (~0.22 bit/s)
3	200 (~0.44 bit/s)
4	500 (~1.11 bit/s)
5	1,000 (~2.2 bit/s)
6	2,000 (~4.4 bit/s)
7	5,000 (~11.1 bit/s)
8	10,000 (~22 bit/s)
9	20,000 (~44 bit/s)
10	50,000 (~111 bit/s)
11	100,000 (~0.22 kbit/s)
12	200,000 (~0.44 kbit/s)
13	500,000 (~1.11 kbit/s)
14	1,000,000 (~2.2 kbit/s)
15	2,000,000 (~4.4 kbit/s)
16	5,000,000 (~11.1 kbit/s)
17	10,000,000 (~22 kbit/s)
18	20,000,000 (~44 kbit/s)
19	50,000,000 (~111 kbit/s)

Table 8: Mean throughput classes for GPRS.

2.8   GPRS typical procedures

Typical GPRS procedures are GPRS Attach, Routing Area Update and PDP context activation.

GPRS Attach (similar to IMSI attach of the classical GSM registration procedure at phone turn on) includes an authentication phase to check the subscriber rights and the allocation of the *temporary identity* (TLLI). After GPRS Attach, the location of the MS is tracked, the communication between MS and SGSN is secured, charging information is collected, the SGSN knows what the subscriber is allowed to do, HLR knows the location of the MS with the accuracy of the SGSN level.

A PDP context is an association between the MS and the GPRS network, and must be created before any data transfer occurs. PDP context functions are network level functions, used to bind an MS to a PDP addresses. Various PDP contexts can be activated, each of them having different QoS attributes. An MS may have several active PDP contexts, e.g., each with its own IP address.

Typically, PDP addresses are dynamic: the HPLMN or VPLMN (*Home/Visited Public Land Mobile Network*) operator assigns a PDP address to the MS only when a PDP context is activated. For Web browsing, typically an Internet class C address is dynamically assigned (note that a class C IP address may be used to support up to 254 hosts).

For each session, information sets called PDP contexts, are created and stored in the MS, the SGSN and GGSN. Each subscribed PDP contains:
-   PDP type (e.g., IPv4, X.25)
-   PDP address (e.g., 128.200.192.64, *.*.*.*)
-   Subscribed QoS profile
-   *Access Point Name* (APN), i.e., the access point address of the GGSN that is used as gateway (GW) to an external PDN
-   VPLMN address allowed.

PDP context activation is initiated by the MS (i.e., a request from an application on the terminal, for instance through the PPP protocol). The procedures involved in PDP context activation are described below.

The MS sends an "Activate PDP Context Request" to SGSN containing the following data:
- PDP Type (IP)
- PDP Address (empty = dynamic)
- QoS
- APN
- Other options

Note that the APN refers to the external network the subscriber wants to use (i.e., physical/logical interface in GGSN).

Then, the SGSN checks subscription data:
- APN
- Dynamic / static IP address.

The SGSN gets the GGSN IP address from a DNS in the GPRS backbone (SGSN-GGSN) IP network; thus the APN is mapped to the GGSN IP address.

The SGSN sends "Create PDP Context Request" to GGSN:
- PDP Type (e.g. IP)
- PDP Address (if empty => dynamic address)
- QoS
- APN
- Other options.

GPRS Release 1998 PDP context management specifies a tight connection between one PDP address and one application (not a flow) [26]. Hence, the PDP context contains the agreed QoS requirements of a particular application. Only one PDP context is assigned to one PDP address. However, more applications that require identical QoS can use the same PDP context and the same PDP address.

UMTS/GPRS Release '99 standards allow for the support of multiple PDP contexts with the same IP address [26],[27]:

- Possibility to negotiate different QoS for different applications (e.g., real-time QoS suitable for multimedia applications and non-real-time QoS suitable for file transfers and WWW browsing);

- A filter (*Traffic Flow Template*) is used in the GGSN to· map downlink data packets to the right PDP context.

Let us consider an MS that has already an active PDP context with the GPRS network. When this terminal has a packet to transmit, it has to set a communication context on the air interface by initiating a TBF establishment (see the Part II of this book, sub-Sections 3.4.1 and 3.4.5).

2.8.1   GPRS tunneling protocol architecture

GPRS supports the access to IP and X.25 networks. Moreover, IP is also the network protocol of the GPRS backbone (SGSN-GGSN).

A DNS is used for mapping logical names to IP addresses in the GPRS backbone network. In particular, APN to GGSN's IP address, Routing Area ID to SGSN's IP address. Each SGSN shall have an access to the DNS functionality in order to perform this mapping (see Fig. 19).

The GPRS backbone and the GPRS subscribers use different IP address spaces.

Mobile IP can be used on top of GPRS; GPRS just gives an IP address to the user.

The GPRS backbone carries IP traffic of the MS in a GPRS tunnel. The *GPRS Tunnel Protocol* (GTP) allows the tunneling of multi-protocol packets between various GSNs. A GTP tunnel is characterized by a TID (*Tunnel IDentifier*). In Release '99, *Tunnel Endpoint Identifier* (TEID) replaces TID [28],[29].

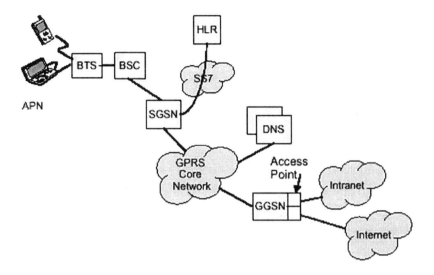

Fig. 19: GPRS network elements to access the Internet.

In order to reach their final destination, data coming from external networks are tunneled twice: into GTP packets in the GPRS core network (GGSN-SGSN) and into LLC frames in the access network (SGSN-MS) [30]. The two-level tunneling mechanism corresponds to a two-level mobility management: LLC "tunnels" (or virtual circuits) correspond to small area mobility (change of SGSN under the same GGSN), while GTP tunnels correspond to wide area mobility (change of GGSN).

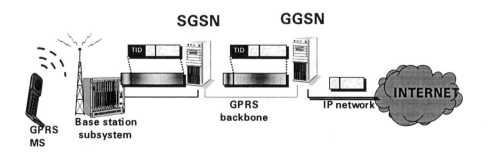

Fig. 20: GPRS tunneling approach for IP traffic.

Let us consider that there is a host connected to the Internet that wants to send information to the MS. When the host transmits information, it will use the PDP address that the HPLMN of the mobile user has assigned to the MS. As a result, the packet will be routed to the GGSN of the user home network. When the GGSN receives the packet it makes a query to the HLR to find the address of the SGSN that is currently serving the user. The GGSN then encapsulates the incoming packet and tunnels it to the appropriate SGSN. Once the SGSN receives the tunneled packet, it looks up the location of the user in its location registry. At that point the SGSN de-capsulates the packet and delivers it to the mobile user. The behaviors of SGSN and GGSN are analogous respectively to those of foreign and home agents in Mobile IP networks.

2.8.2   GPRS protocol stack

This Section briefly describes the GPRS protocol stack on the air interface, referring to both Figs. 16 and 21.

IP packets and all relevant transport protocol headers are forwarded to the *Subnetwork Dependent Convergence Protocol* (SNDCP) layer that formats the packets for transmission over the GPRS network [31]. The SNDCP also carries out header compression, multiplexing of data coming from different sources and carries network layer protocol data units in a transparent way.

LLC provides a logical link between the MS and the SGSN. Its functions include ciphering, flow control and sequence control. In acknowledged mode, LLC provides detection and recovery of transmission errors.

As long as the network packet (datagram) size does not exceed the maximum LLC frame size (1520 bytes), each IP packet is mapped onto a single LLC frame (SDU). The LLC frames are passed onto the RLC/MAC layer where they are segmented in fixed-length RLC/MAC blocks (PDUs). The MAC protocol then carries out procedures that allow multiple MS to share a common transmission medium that may consist of several physical channels. The MAC protocol allows one MS to use many time slots, but also many MSs to share the same time slot.

RLC can be operated in acknowledged or unacknowledged mode.

RLC/MAC blocks are divided in GSM bursts to be transmitted on the air interface, where the physical layer is responsible for error correction and detection. Rectangular interleaving of radio blocks and procedures for detecting physical link congestion are also carried out in this layer. GPRS data is transmitted over the logical channel named *Packet Data Traffic CHannel* (PDTCH). This channel is mapped onto one PDCH resource (= slot).

Base Station System GPRS Protocol (BSSGP) transmits QoS-related information between BSS and SGSN [32]. It specifies a flow control for the downlink direction. The BSS on the basis of the buffer occupancy can change the setting of shaper parameters in the SGSN to regulate the MS flows or aggregate flows directed to a radio cell.

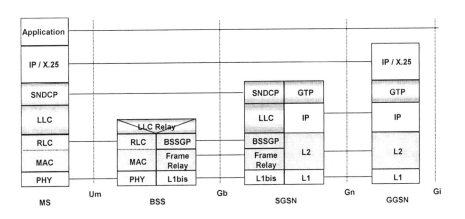

Fig. 21: GPRS protocol architecture at different interfaces.

2.9   GPRS services

A significant advantage of GPRS compared to a circuit-switched mobile technology, e.g., GSM, is that it provides the possibility to a user to be online connected to the network without being charged for the time it remains in that situation provided that no traffic is generated.

As GSM, GPRS supports SMS. Moreover, GPRS has two broad categories of bearer services: *Point-to-Point* (PTP) and *Point-to-Multi-point* (PTM).

The forwarding process of PTM can be accomplished in two different ways. It can be either multicast, i.e., sent to all receivers located in a geographical area, referred to as PTM-M (*PTM-Multicast*), IP-M (*Internet Protocol Multicast*) or forwarded to a predefined group (mainly independent of geographical location), referred to as PTM-G (*PTM-Group*).

GPRS also supports variable QoS. QoS profiles can be negotiated between the mobile user and the network for each session, depending on the QoS demand and the current available resources. QoS profiles can be defined as described in the previous Section 2.7: service precedence, reliability, delay, and throughput.

Chapter 3: 3G mobile systems

Third-generation (3G) wireless communications is the evolution of second-generation (2G) mobile systems towards increased data rates as well as a wide range of advanced services (from the traditional voice and paging services to interactive multimedia including high-quality teleconferencing and high speed Internet access) [33].

The *International Telecommunication Union* (ITU) began its activities dealing with future mobile communication systems in 1985 under the name *Future Public Land Mobile Telecommunication System* (FPLMTS). ITU-R Task Group 8/1 was in charge of identifying FPLMTS needs. Subsequently, ITU-R changed the name of FPLMTS to *International Mobile Telecommunications after the year 2000* (IMT-2000), whose air interface standardization studies began in 1997.

Some requirements for 3G technologies are:
- More efficient use of the available spectrum,
- Support for a wide variety of mobile equipment,
- Flexible introduction of new services and technologies,
- Voice quality comparable to that of the *Public Switched Telephone Network* (PSTN),
- A data rate of 144 kbit/s for users moving fast over large areas,
- A data rate of 384 kbit/s for pedestrians or slow moving users over small areas,
- Support for 2.048 Mbit/s operation for office use,
- Support for both packet-switched and circuit-switched data services,
- An adaptive radio interface suited to the highly asymmetric nature of most Internet communications,
- A much greater bandwidth for downlink than for uplink.

In 1992, the ITU *World Administrative Radio Conference* (WARC-92) identified a worldwide frequency band around 2 GHz for the deployment of third-generation mobile communication systems (see Fig. 22). In particular, frequency bands 1885-2025 MHz and 2110-

2200 MHz were allocated [34]. Within these bands, 1980-2010 MHz and 2170-2200 MHz were destined for the satellite component of 3G systems.

Europe and Japan have decided to implement the terrestrial part of 3G systems based on *Frequency Division Duplexing* (FDD) in the paired bands 1920 - 1980 MHz and 2110 - 2170 MHz. Europe has also destined the unpaired bands 1900 - 1920 MHz and 2010 - 2025 MHz for the *Time Division Duplexing* (TDD) transmission mode. More details on European bandwidth allocations are shown in Fig. 23.

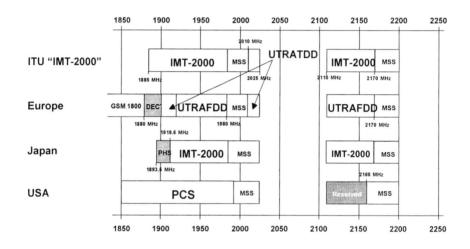

Fig. 22: Frequency allocations for 3G systems in the World.

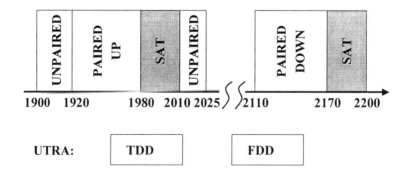

Fig. 23: European frequency allocations for 3G air interfaces.

Within IMT-2000, the 3G standardization process was done in parallel by the following standardization bodies throughout the World:

- *China Wireless Telecommunication Standard group* (CWTS) in China,
- *European Telecommunications Standards Institute* (ETSI) in Europe,
- *Telecommunications Industry Association* (TIA) in the United States,
- T1 Standardization Committee in the United States,
- *Association of Radio Industries and Businesses* (ARIB) in Japan,
- *Telecommunications Technology Association* (TTA) in South Korea,
- *Telecommunications Technology Committee* (TTC) in Japan.

Each standardization body had to submit its own proposal to ITU-R by June 1998 complying with ITU-R Recommendation M.1225.

The definition of 3G-radio interface characteristics has been the subject of major research efforts for more than ten years. For example, ETSI has coordinated and funded a large number of technology development projects both under the RACE (*Research on Advanced Communications for Europe*) [35] and ACTS (*Advanced Communications Technologies and Services*) programs. One output of these ETSI programs was the detailed definition of different radio interfaces using (see Fig. 24):

- *Wideband CDMA* (WCDMA) - concept group α,
- *Wideband TDMA* - concept group γ,
- *Wideband Time-Division Code-Division* (TD/CDMA) - concept group δ,
- *Orthogonal Frequency Division Multiple Access* (OFDMA) technologies - concept group β,
- *Opportunity Driven Multiple Access* (ODMA) - concept group ε.

Detailed comparisons of these radio interface technologies were made in order to obtain a consensus on the proposals to be submitted to ITU-R by June 1998. Towards this end, the following objectives were considered:

- Low-cost terminal,
- Harmonization with GSM,
- FDD / TDD dual-mode operation,
- Fit into 2 - 5 MHz spectrum allocation.

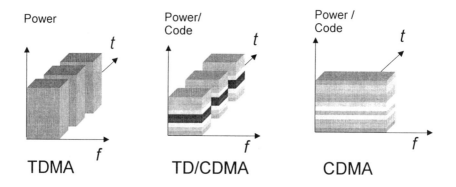

Fig. 24: Air interface techniques.

A decision was taken in January 1998 that the European recommendation to the ITU for the 3G air interface (i.e., *Universal Mobile Telecommunications System* -UMTS- *Terrestrial Radio Access*, UTRA) would be WCDMA for the frequency division duplex spectrum (paired band; see Fig. 23) and TD/CDMA for the time division spectrum (unpaired band; see Fig. 23). These schemes were respectively denoted as UTRA-FDD and UTRA-TDD (see Fig. 25).

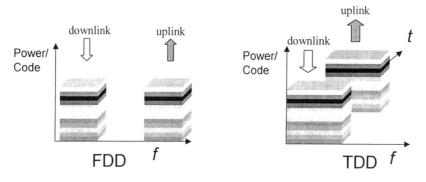

Fig. 25: FDD and TDD options for UTRA.

In June 1998, ETSI submitted the UTRA specifications to ITU-R to be included into the IMT-2000 standards family (of course, UTRA specifications are still improved and refined in subsequent releases). In June 1998, 10 terrestrial and 6 satellite different air interface proposals were submitted to ITU-R. Both time-division (TDMA) and code-division (CDMA) multiple access technologies were considered although the majority of these proposals was based on wideband CDMA (WCDMA) technology.

Focusing on terrestrial systems, basically two standards emerged: the European WCDMA UTRA-FDD proposal and the US cdma2000 proposal. An *Intellectual Property Right* (IPR) war started on CDMA patents between the Sweden Ericsson and the US Qualcomm Inc.

In October 1998, OHG (*Operator Hamonisation Group*) started its activities with the aim to achieve some form of interworking between WCDMA and cdma2000 under the name of *Global Third Generation* (G3G). The crisis between Ericsson and Qualcomm was solved in March 1999 when the parties signed an agreement on IPR cross licensing.

In December 1998 several organizations bodies agreed to launch the *3rd Generation Partnership Project* (3GPP) whose objective was the adoption of WCDMA on the air interface and MAP protocols in the core network [36]. Presently 3GPP brings together ARIB, CWTS, ETSI, T1, TTA and TTC. 3GPP standardizes service capabilities that consist of bearers defined by QoS parameters and the mechanisms needed to realize services (including various network elements, the communication between them and the storage of associated data). In order to achieve global harmonization, 3GPP has modified some parameters of the UTRA proposal (e.g., the chip-rate has changed from the original value of 4.096 to 3.840 Mchips/s; a new downlink pilot structure has been adopted; the asynchronous/synchronous base station operation has been modified).

Contemporaneously, the *3rd Generation Partnership Project 2* (3GPP2) was settled with the aim of promoting the standardization of a 3G mobile system based on the evolving ANSI-41 core network and on the cdma2000 air interface [37]. 3GPP2 is a collaborative effort between ARIB, CWTS, TIA, TTA, and TTC.

In March 1999 the final decision was made by ITU to select a single access scheme, wideband CDMA, with three different modes (see Fig. 26): *Direct Sequence* (DS) FDD, based on ETSI and ARIB WCDMA FDD proposals; *MultiCarrier* (MC) FDD, based on TIA cdma2000 proposal; DS TDD, based on ETSI (UTRA-TDD) proposal. In addition to this, two TDMA air interfaces were included as evolutions of 2G wireless systems.

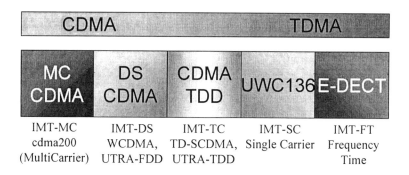

Fig. 26: IMT-2000 air interfaces options.

3GPP (UTRA-FDD, WCDMA) and cdma2000 are both wideband CDMA systems with some similarities, but also with many implementation differences. For example, the user bit-rate is spread to the final rate of either 1.2288 Mchip/s for cdma2000 (the same rate used by the IS-95A standard with a 3x version proposed as a future upgrade) or 3.84 Mchip/s for 3GPP. Moreover, there are considerable differences in the coding, synchronization, base station identification methodologies and network protocols. WCDMA is a new system, designed only for use in new spectrum (IMT-2000 band). It will require new equipment installation by the network operators. Whereas, cdma2000 is an improvement of IS-95, so that existing equipment can be upgraded to support the new system. Cdma2000 has been designed to share the same frequency in each sector of each cell of IS-95, but higher efficiency is achieved.

In Europe, many GSM operators will migrate to WCDMA for increased data capabilities. Korea is evaluating WCDMA along with cdma2000; at least one of the three 3G licenses will be based on cdma2000 technology. Note that Korea has the highest concentration of IS-95 users in the World. They will continue in this technology with the rollout of cdma2000. China will consider several options including TD-SCDMA, WCDMA and cdma2000. In the Americas, existing IS-95 operators will migrate to the cdma2000 family of standards for increased capacity: 1xRTT, 1xEVDO (data only) and 1xEV-DV (data/voice). As for Japan, a 3G system named FOMA (*Freedom Of Multimedia Access*) has been already deployed by NTT DoCoMo and started its operations from October 2001. FOMA is based on WCDMA technology and allows the provision of voice and high-speed data communications for mobile users. The maximum downlink speed is 384 kbit/s for the delivery of high-quality video transmissions and high-bit-rate i-mode services.

3G systems are based on a hierarchical cell structure (see Fig. 27), designed to support increasing capacity with lower bit-rates as the user moves from the downtown area (with micro-cellular coverage) to a suburban area (with macro-cell coverage).

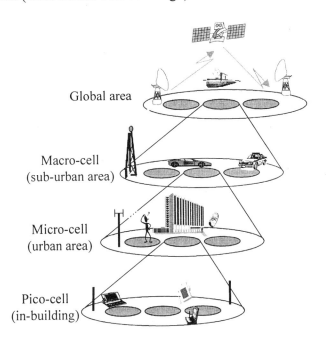

Fig. 27: Hierarchical 3G system architecture with related mobility degrees. User bit-rate will be up to 2 Mbit/s in pico-cells, 384 kbit/s in micro-cells and 144 kbit/s in terrestrial and satellite macro-cells.

3.1   UMTS traffic classes

The new paradigm behind 3G systems is the widespread provision of multimedia services and applications to users while on the move. The supported bit-rates will be at least 144 kbit/s for rural area, 384 kbit/s for urban/suburban area and 2 Mbit/s for indoor/low range outdoor environment.

Four different traffic classes have been identified by 3GPP technical specifications [38]. They are listed below with the conversational class having the most stringent delay constraint and the background class having the loosest delay constraint.

-   *Conversational*: real-time-oriented (voice),

-   *Streaming*: controlled delay variations (streaming video),

-   *Interactive*: low data loss and time constrained (Web, chat, games),

-   *Background*: low data loss with no time guarantees (e-mail, SMS).

Conversational and streaming classes are mainly intended to carry real-time traffic flows. The main difference between them is how delay sensitive the traffic is. Conversational real-time services, like video telephony, are the most delay-sensitive applications. Streaming traffic class may tolerate higher transfer delays.

Interactive and Background classes are used by traditional Internet applications. Due to looser delay requirements, compared to conversational and streaming classes, both provide better error rate performance by means of channel coding and retransmissions. The Interactive class is mainly used by interactive (transactional) applications (e.g., interactive applications via Web), while the Background class is meant for download of e-mails. Interactive traffic has a higher priority than Background traffic. The detailed characteristics of these different traffic classes are described below with reference to Table 9.

Conversational class

This is the typical traffic class for telephony speech. In 3G systems, the conversational class will be used for Voice over IP (VoIP) and for video conferencing. Real-time conversation is always performed between peers (or groups) of human end-users; hence, the required characteristics are strictly related to human perception. It is therefore necessary that the transfer time be low due to the conversational nature of the scheme and that the time variations between information entities of the same stream be preserved.

Streaming class

This traffic class is for applications where a user is looking at (listening to) real-time video (audio). Streaming requires that time relations between information entities (i.e., samples, packets) within a flow be preserved; however, there is no requirement on low transfer delays. Only the delay variations of the end-to-end flow must be limited to preserve the time relation between information entities of the stream.

Interactive class

This traffic class is used for applications where a machine or a human is on line requesting data from a remote equipment (e.g., a server). Examples of human interaction with a remote equipment are: Web browsing, data base retrieval, server access. Examples of machines interaction with a remote equipment are: polling for periodic measurement records and automatic data base inquiries. Interactive traffic is characterized by a request-response pattern: at the destination there is an entity expecting the response within a certain time. Therefore, a key attribute is represented by the round trip delay. Another important requirement is the integrity of the transferred information.

Background class

This is the traffic class adopted by a computer for background data transmissions (e.g., background delivery of e-mail, SMS). There is no delay requirement for background traffic, since the destination is not expecting the data within a certain time. However, the content of the packets must be transferred with low error rate.

Traffic Class	Conversational	Streaming	Interactive	Background
Fundamental characteristics	Preserve time variation between information entities of the stream; stringent and low delay	Preserve time variation between information entities of the stream	Request-response pattern; preserve payload content	No delay constraint; preserve payload content
Application examples	Voice, Video, Interactive games	Streaming audio, Streaming video	Web browsing, Email, Ftp, Database retrieval	Background download of e-mail, Backdrop delivery of e-mail, SMS

Table 9: UMTS traffic classes.

The requirements for the different traffic classes are given in [39]. Some details are shown in Table 10.

Bearer	Conversational	Streaming	Interactive	Background
UMTS Bearer (RAB + CN Bearer)	150ms (from 23.107)	250ms (from 23.107)	400 ms (back of envelope calculation based on 10x1500 bytes packets per web page)	1000ms (guess)
Radio Access Bearer (RAB) (UTRAN + Iu)	120ms (from 23.107)	200ms (80% of UMTS bearer)	320ms (80% of UMTS bearer)	800ms (80% of UMTS bearer)
CN Bearer (from MSC or SGSN to Gateway)	30ms	50ms	80 ms	200ms
Iu Bearer	24 ms (20% of RAB)	40ms (20% of RAB)	64 ms (20% of RAB)	160ms (20% of RAB)
Radio Bearer	96 ms (80% of RAB)	160ms (80% of RAB)	256 ms (80% of RAB)	640ms (80% of RAB)

Table 10: UMTS QoS requirements for the different traffic bearers (see the following Section 3.3).

3.2 UMTS architecture description

The UMTS network consists of a *Radio Access Network* (*UMTS Terrestrial Radio Access Network*, UTRAN) and a *Core Network* (CN). UTRAN is formed by a hierarchical *Radio Network Sub-system* (RNS), whose elements are: *Radio Network Controller* (RNC), Node B and *User Equipment* (UE) [40]. UE is the combination of the *Mobile Equipment* (ME) and the *Universal Subscriber Identity Module* (USIM).

The UMTS Release '99 reference architecture is shown in Fig. 28, with further details in Fig. 29. The UMTS Release '99 architecture inherits a lot from the GSM-GPRS system on the CN side. The MSC has basically very similar functions in both GSM and UMTS. Whereas, packet data traffic uses the GPRS network with its related SGSNs and GGSNs.

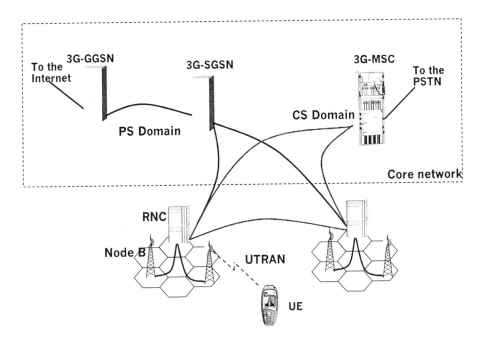

Fig. 28: UMTS - Release '99 network architecture with highlighted *Packet-Switched* (PS) and *Circuit-Switched* (CS) domains.

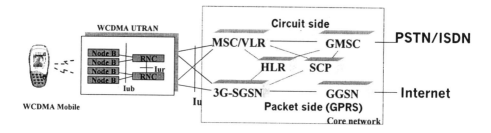

Fig. 29: UMTS - Release '99 detailed architecture.

UTRAN consists of a set of RNSs connected to the CN through the Iu interface (see Fig. 30). Node Bs are connected to UE by means of the Uu interface (air interface).

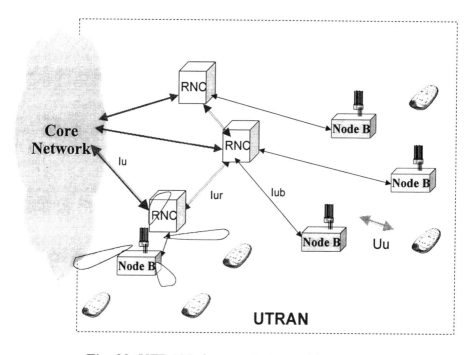

Fig. 30: UTRAN characteristics and interfaces.

The Iu interface itself is separated into the Iu-CS interface between the RNC and the circuit-switched domain of the CN, and the Iu-PS interface between the RNC and the packet-switched domain of the CN (see Fig. 28).

An RNS consists of an RNC and one or more Node Bs. A Node B is connected to the RNC through the Iub interface.

Each RNC is responsible for the management of the radio resources of its set of cells; each Node B supports one or more cells. The RNC is responsible for handover decisions that require signaling to the UE.

With WCDMA macro-diversity combining can be employed during handoff to exploit the signal of two cells (soft handover). RNCs are interconnected using the Iur interface. Such interconnection enables the RNCs to keep handoffs within the UTRAN.

UMTS - Release '99 is designed to allow the coexistence of UTRAN with the classical GSM-GPRS base station sub-system, so that both access networks share the same CN facilities (see Fig. 31).

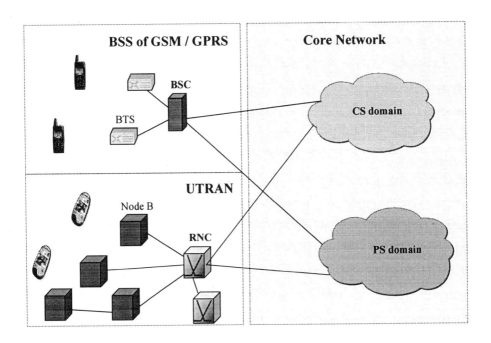

Fig. 31: Coexistence of GSM and UTRAN in UMTS - Release '99.

The UMTS protocol architecture envisages a stack divided into the following two parts:

- **User plane** for data transport: circuit-switched domain and packet-switched domain,

- **Control plane** for signaling.

In Release '99, *Asynchronous Transfer Mode* (ATM) is used as the transport technology in UTRAN due to its capabilities to support heterogeneous traffic classes with QoS guarantee. In particular, ATM is used at the interface between Node Bs and RNCs, between RNCs, and between RNCs and the CN [41]. Conversion of ATM to the circuit-switched *Time Division Multiplexed* (TDM) technology (if used to switch the voice paths in the core network) can be performed by the MSC or by a gateway function between the RNC and the MSC. Both circuit-switched and packet-switched services are carried by ATM cells, using appropriate *ATM Adaptation Layer* (AAL) protocols [42]. In particular, AAL2 is used for voice bearer circuits[3] (user plane); whereas, AAL5 is used for Internet packet data traffic and for signaling (control plane of both circuit and packet switched bearers). Fig. 32 shows the protocol architecture for Internet access (user plane) at the different interfaces [43],[44].

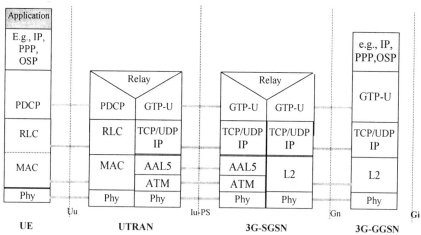

Fig. 32: Interfaces of the user plane for packet-switched traffics.

---

[3] AAL2 is used to transport compressed speech efficiently and within acceptable delay limits in ATM networks.

Whereas, the protocol stack for voice traffic at the Iu interface is shown in Fig. 33.

AAL-2 SAR SSCS (I.366.1)
AAL2 (I.363.2)
ATM

Fig. 33: Iu-PS Interface protocol stack (user plane).

Due to the success and widespread diffusion of the Internet, the IP protocol has been considered for adoption in the 3G networks instead of the ATM layer. This evolution is envisaged to cope with the need of an efficient mobile access to the Internet (IP over ATM is an inefficient solution). The success of *IP Telephony* (IPT) and the recent standardization of the H.323 umbrella of standards are other important reasons why IP-based transport is a promising option for cellular networks. Note that an IP-based control plane enables dynamic configuration of network elements and topologies. For instance, an RNC can be added to UTRAN in a "plug and play" way.

Therefore, the following evolution phases are envisaged for UMTS towards the IP support:

- *Release 4*, which includes the migration of the Release '99 circuit-switched domain CN to an IP transport,

- *Release 5*, which envisages the support of conversational and interactive multimedia services on an end-to-end IP transport provided by an enhanced GPRS network (PS domain).

At present, Release 6 is in the standardization phase. It has to be completed by 2003 with new potentialities for 3G services.

Releases 4 and 5 were previously known singly as Release 2000.

The evolved UMTS network architecture contains new blocks, as follows (see Fig. 34):

- MGW (*Media Gateway*): conversion between PS and CS on the data plane,

- CSCF (*Call State Control Function*): mobility management, call control,

- MGCF (*Media Gateway Control Function*): control of the MGW on the basis of the H.248 protocol and conversion between CS and PS,

- SGW (*Signalling Gateway*): translation between SS7 and Sigtran at transport level.

Release 5 introduces new features:

- Provisioning of IP-based multimedia services as an extension of PS services,

- Bearer independence,

- Separation of transport and control (e.g., the classical MSC is split into MSC server and MGW),

- *Internet Multimedia* (IM) sub-system that comprises all CN elements for the provision of IP multimedia services (e.g., CSCF, MGCF),

- HSS (*Home Subscriber Server*) that contains user and terminal profiles.

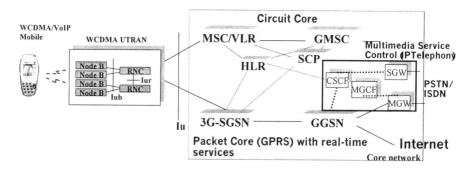

Fig. 34: Release 4/5 general network architecture.

A detailed Release 5 network architecture is shown in Fig. 35.

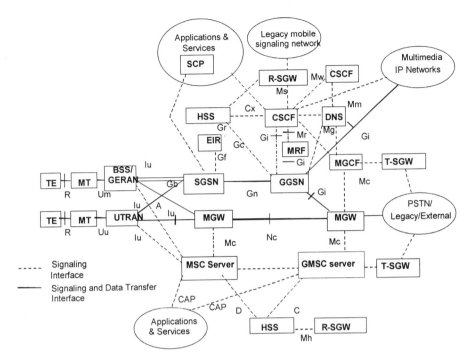

Fig. 35: Release 5 detailed network architecture.

In the Release 5, a new core network domain, the IM core network sub-system is introduced for the control of voice and multimedia calls and sessions and the interconnection to other networks, such as PSTN and other UMTS networks. This is a PS domain that is added to the GPRS core network (i.e., SGSN-GGSN).

The main reasons for the introduction of the IM domain are to enable new services and to reduce costs. The IM domain uses IP and the other protocols standardized by the *Internet Engineering Task Force* (IETF) as interfaces to component 'building blocks' of the Release 5 network. The IP network of the IM sub-system is enabled to provide the QoS needed for voice and multimedia services.

Examples of the services supported in Release 5 by the IM domain are:

- Voice telephony,
- Real-time interactive games,
- Video-telephony,

- Instant messaging,
- Emergency calls,
- Multimedia conferencing.

These services generally require a *conversational session* between two or more parties. The real-time aspects of the service can be described in terms of the QoS of the transport (e.g., transmission delay or packet jitter) and of the session control (e.g., time to establish a session). To meet the degree of interactivity needed by these services, the GPRS IP core network provides QoS levels by employing, for example, DiffServ [45],[46] (more details are given later in this Section). Moreover, IP version 6 (IPv6) [47],[48] has been recommended as the protocol to be used for the IM domain, since this has a number of features that are beneficial to UMTS networks (such as a large address space, support for packet prioritization and easier manageability).

IPv6 is a new version of the *Internet Protocol*, designed as a successor to IP version 4 (IPv4) [49], the predominant protocol in use today in the Internet. The changes from IPv4 to IPv6 are primarily in the following areas: larger addressing capabilities, header format simplification; flow labeling capability; consolidated authentication/privacy capabilities. The maximum number of 4 billion public addresses allowed by Ipv4 will be inadequate in the future. Therefore, IP version 6 is the solution to expand the address space: IPv6 has 128-bit long addresses (about 3.4×10^{38} different possible configurations).

The following Fig. 36 presents the user plane protocol stack for 3G PS conversational applications explaining the transport of different media types [50].

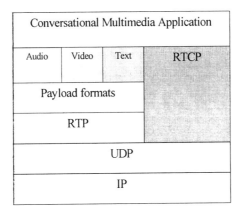

Fig. 36: User plane protocol stack for PS conversational traffic.

The IM domain has four important roles in meeting the requirements of services:

- It enables users and applications to control sessions and calls between multiple parties (e.g., to establish, to maintain, to modify and to terminate calls).

- IM controls and supports network resources (such as MGWs and GGSNs, *Multimedia Resource Functions*, MRF, and the core IP network) to provide the functionality, security and quality required for the call.

- IM provides for registration of users on the 'home' and 'roamed to' networks, so that users may access their own services from any UMTS network.

- IM generates *Call Detail Records* (CDRs), containing information on time, duration, volume of data sent/received, and the call participants; CDRs together with records from the GPRS network on the data volumes transmitted and received are used for charging purposes.

The signaling protocol for registration and call control in the IM domain is based on the *Session Initiation Protocol* (SIP) [51]. To send and receive SIP messages over the GPRS network, the user equipment must establish a bi-directional packet data session with the IM domain for the signaling path. This requires PDP context activation, a common GPRS procedure for establishing an IP data path between the UE and the Internet. SIP only supports the call control procedures for the establishment of the speech path. The allocation of the actual bandwidth and the QoS provision for the IP transport of speech packets over UTRAN and GPRS networks is requested as an additional PDP context, by using the GPRS protocols.

IP networks commonly provide best-effort data delivery by default. Hence, two QoS approaches have been introduced at the IP level by the IETF; in particular, *Integrated Services* (IntServ) and *Differentiated Services* (DiffServ). Correspondingly, two QoS management schemes are considered, as described below.

- **Resource Reservation**. Network resources are allotted according to an application QoS request, and are subject to bandwidth management policy. *Resource Reservation Protocol* (RSVP) in the IntServ architecture provides signaling for implementing this QoS approach. IntServ entails admission control, policing, traffic shaping and scheduling of packets.

- **Prioritization**. Network traffic is classified and network resources are assigned according to bandwidth management criteria. Network elements give preferential treatment to traffics with higher QoS requirements. This strategy is implemented by DiffServ that aims to guarantee QoS levels on a per-flow (i.e., traffic class) basis. DiffServ networks classify packets into one of a small number of aggregated flows or *classes*, based on the *DiffServ CodePoint* (DSCP) in the IP packet header (*Type of Service* byte in the IPv4 header and *Traffic Class* byte in the IPv6 header). This is known as *Behavior Aggregate* (BA) classification. At each DiffServ router, packets are subjected to a *Per-Hop Behavior* (PHB) that is invoked by the DSCP. Basically, three different traffic classes (PBHs) can be considered:
    - *Expedited Forwarding* (EF) for guaranteed bandwidth, low and bounded jitter, low delay, low packet losses;
    - *Assured Forwarding* (AF) for services requiring minimum assured bandwidth (additional bandwidth, if available) with possible packet dropping above the agreed data rate in case of resource shortage;
    - *Best-Effort* for common Internet traffic.

3.3   UTRAN resources

UTRAN resources at different protocol layers are (see Fig. 37): logical channels, transport channels and physical channels.

Logical channels (LoCH) are information streams dedicated to the transfer of specific types of information. Logical channels can be broadly classified into two groups: control channels (for the transfer of control plane information) and traffic channels (for the transfer of user plane information). MAC is responsible for mapping logical channels on transport channels.

Transport channels (TrCH) describe how data is transported over the radio interface. Transport channels are the services offered by the physical layer to the MAC layer. They are managed at the RNC level. A general classification of transport channels is into two groups: common transport channels and dedicated transport channels.

- *Common channels* (CCHs)
    - *Broadcast channels* (BCH) (DL)
    - *Forward-Access Channels* (FACH) (DL)
    - *Paging Channel* (PCH) (DL)
    - *Random-Access Channel* (RACH) (UL)/Slotted Aloha
    - *Common Packet Channel* (CPCH) (UL)
    - *Downlink Shared Channel* (DSCH) (DL)/Multicast

- *Dedicated Channels* (DCH) (UL/DL)

Each transport channel has a transport format, indicating how data is transmitted over the radio interface. Transport channels are mapped on physical channels.

Physical channels are the radio physical resources between UEs and Node Bs. Physical channels allow different kinds of bandwidth allocated for different purposes. Physical channels are detailed in the following Sections on physical layer description.

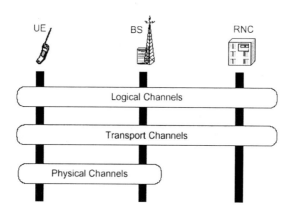

Fig. 37: Channels in UTRAN.

UMTS is based on the *stratum* concept that entails the grouping of protocols related to one aspect of the services provided by one or several domains (see Fig. 38).

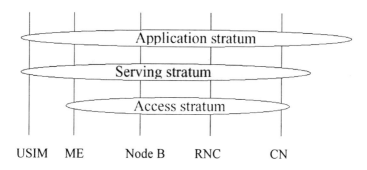

Fig. 38: UMTS stratum concept.

Bearer services provide the capability for information transfer between network points. Bearer is a capability of defined capacity, delay and bit error rate, etc. It is a flexible concept designating a kind of bit pipe at a certain network level (*stratum concept*), between given network entities, with certain QoS attributes and capacity (see Fig. 39).

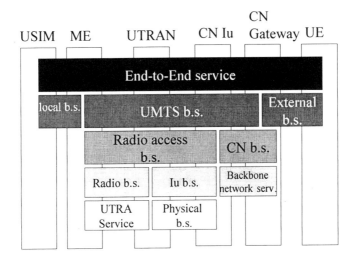

Fig. 39: Bearer services (b.s.) architecture at the different UMTS interfaces [52].

Bearers are characterized by a set of QoS requirements as:

- *Maximum transfer delay*: this is the time between the request to transfer the information at one access point to its delivery at the other access point.
- *Delay variation* of the information received over the bearer (it has to be controlled to support real-time services).
- *Bit Error Rate* (BER), which is the ratio between incorrect and total transferred information bits.
- *Data rate*, the amount of data transferred between two access points.

UTRA radio bearers are [53]:

- 8 kbit/s bearer for circuit-switched voice traffic with low delays (20 ms) and not stringent requirements on quality (BER $< 10^{-3}$).
- LDD (*Low Delay Data*) bearer service for circuit-switched data traffic (at 144, 384, 2048 kbit/s) with low delays (50 ms) and high quality (BER $< 10^{-6}$).

- LCD (*Long Constrained Delay*) data bearer service for data traffic (at 144, 384, 2048 kbit/s) with long delay (300 ms) and high quality (BER < 10^{-6}).

- UDD (*Unconstrained Delay Data*) bearer service for connectionless data packet traffic with no specific requirement on delay but with high quality (BER < 10^{-8}).

These bearers are according to ITU-T Recommendation I.362 that define four service classes:

- *Class A* for connection-oriented constant bit-rate real-time traffic,

- *Class B* for connection-oriented variable bit-rate real-time traffic,

- *Class C* for delay unconstrained, connection-oriented traffic,

- *Class D* for delay unconstrained, connectionless traffic.

3.4   UMTS air interface: characteristics of the physical layer

In CDMA, all communicating units transmit at the same time and in the same frequency. Multiple access is achieved by assigning each user (or channel) a distinct code.

The European Project *Future Radio wideband Multiple accEss System* (FRAMES, 1995-1998) within the ACTS program can be technically considered as the 'father' of UTRA. Based on the previous work carried out in the RACE II projects CoDiT (on CDMA) and ADTMA (on TDMA), FRAMES identified two air interface modes (considered in the UTRA definition process described at the beginning of this Chapter):

- FMA1: TDMA with CDMA (practically, an evolution of GSM) able to support joint detection,

- FMA2: WCDMA (evolution of CoDiT concept).

These are the interfaces adopted, with some modifications, by the UMTS standard respectively for unpaired bands with TDD and paired bands with FDD. 3GPP specifications of the 25-series describe all the characteristics of both FDD and TDD modes of UTRA air interface [54]-[79] as well as Iu, Iub and Iur interfaces.

3.4.1 UTRA-FDD physical layer characteristics

The WCDMA chip rate is 3.84 Mchip/s, that corresponds to a modulated carrier band of 5 MHz due to the use of smoothed (root-raised cosine) chip impulses with roll-off factor $r = 0.22$. Operators can deploy multiple carriers, each of them occupies 5 MHz (see Fig. 40). The frame length is 10 ms and each frame is divided into 15 slots (2560 chips/slot at the chip-rate of 3.84 Mchip/s). Each frame conveys a packet per code. The *Processing Gain* (PG) ranges from 256 to 4 in uplink and from 512 to 4 in downlink.

The WCDMA physical layer includes the variable-bit-rate transport channels required for bandwidth-on-demand user applications. These can multiplex several services onto a single connection between the fixed infrastructure and a mobile terminal.

Some of the physical channels do not carry user information. In fact, they are used by the physical layer itself, such as pilot channels (that assist in modulation recovery), a synchronization channel (that lets mobile terminals synchronize to the network) and an acquisition channel (that allows initial connections to mobile terminals).

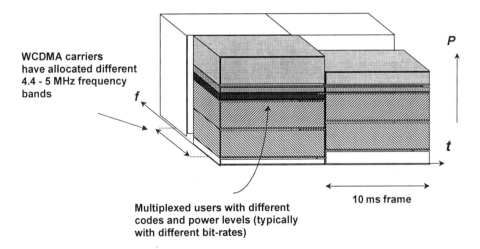

Fig. 40: WCDMA air interface.

WCDMA applies the spreading process in two phases. An initial channelization code spreading is followed by a scrambling code spreading. The channelization code alone determines the spread bandwidth occupied by the radio signal. Whereas, the scrambling code is used to distinguish different UEs at the Node B receiver (uplink) and to distinguish multiple Node Bs in the UE receiver (downlink).

For separating channels from the same source (with different bit-rates), *Orthogonal Variable Spreading Factor* (OVSF) channelization codes are used. OVSF codes are organized according to a tree, as shown in Fig. 41. At each branch, the PG value doubles. During transmissions, the PG value can be updated at each frame (i.e., every 10 ms), thus allowing variable bit-rate transmissions.

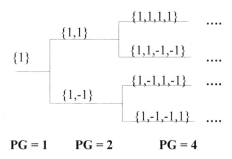

Fig. 41: OVSF code tree (Walsh-Hadamard codes).

OVSF codes of different length are orthogonal if they have not a common path towards the code tree root. Under this constraint, a sender can use more OVSF codes to mix transmissions with different rates, so that they are orthogonal on the air interface. OVSF codes are designed with the symbol $C_{ch,PG,j}$, where PG stands for the processing gain (from 4 to 256 for uplink and from 4 to 512 for downlink) and j denotes the code number, $0 \leq j \leq PG - 1$.

As for the channel coding, three options are supported depending on the BER requirement (see Fig. 42): convolutional coding (code rates 1/2 or 1/3), turbo coding (code rate 1/3) or no channel coding. The selection of the channel coding scheme is decided by upper layers. Bit interleaving is used to randomize transmission errors.

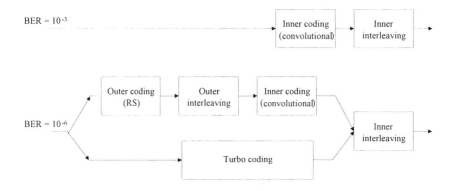

Fig. 42: FEC coding schemes.

Let us first focus on uplink transmissions. There are two dedicated physical and two common physical channels:

- The uplink *Dedicated Physical Data CHannel* (uplink DPDCH) and the uplink *Dedicated Physical Control CHannel* (uplink DPCCH);

- The *Physical Random Access CHannel* (PRACH) and the *Physical Common Packet CHannel* (PCPCH).

The uplink DPDCH is used to carry data generated at layer 2 and above (i.e., the dedicated transport channel, DCH) [63]. There may be zero, one, or several uplink DPDCHs on each layer 1 connection.

The uplink DPCCH is used to carry control information generated at layer 1. Control information consists of known *pilot bits* to support channel estimation for coherent detection, *Transmit Power Control* (TPC) commands for downlink power control, *FeedBack Information* (FBI) to adjust the diversity antenna phase, or phase/amplitude depending on the transmission diversity mode and an optional *Transport-Format Combination Indicator* (TFCI). The TFCI informs the receiver about the instantaneous parameters of the different transport channels multiplexed on the same uplink DPDCH and corresponds to the data transmitted in the same frame.

Before spreading, binary value '0' is mapped to '+1' and binary value '1' is mapped to '-1'.

There are two possible configurations for uplink transmissions: single DPDCH or more parallel DPDCHs (up to 6).

Let us consider the first case: Fig. 43 shows the frame structure of the uplink DPDCH and DPCCH.

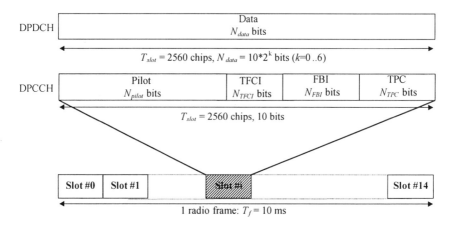

Fig. 43: Frame structure for uplink DPDCH/DPCCH. Each frame of length $T_f = 10$ ms is divided into 15 slots, each of length $T_{slot} = 2560$ chips, corresponding to one power control period.

DPDCH and DPCCH form two distinct bit fluxes that are separately spread by channelization codes. Thus, we obtain phase and quadrature (I and Q) signals that are input to the complex scrambling process. In a slot of 10/15 ms, there are $10*2^k$ bits with $k \in \{0, 1, ...,6\}$ and 2560 chips. Hence, the processing gain PG = $256/(2^k)$ may range from 256 down to 4. The bit-rate as a function of the processing gain, $R_{up}(PG)$, results as: $R_{up}(PG) = (10*2^k)/(10ms/15) = 3840/PG$ kbit/s. The maximum bit-rate for this configuration with one DPDCH is achieved for PG = 4 and is equal to $R_{up}(PG = 4) = 960$ kbit/s. Moreover, the PG value for the DPCCH is always equal to 256, corresponding to a bit-rate of 15 kbit/s. Note that the information bit-rate supported by one DPDCH is obtained from the above $R_{up}(PG)$ value, taking into account the channel coding with rate 1/2 or 1/3. Correspondingly, we have a set of information bit-rate values as: $R_{up\_info}(PG) = (1/2)*3840/PG$ kbit/s or $R_{up\_info}(PG) = (1/3)*3840/PG$ kbit/s. The processing gain on DPDCH (and, hence, the carried bit-rate) may vary on a frame basis.

If the maximum information bit-rate achievable in the configuration with a single DPDCH is not enough, the other possibility is to use 6 DPDCHs and one DPCCH. In this case all the DPDCHs have PG = 4 and the DPCCH has PG = 256. As shown in Fig. 44, 3 DPDCHs are spread by means of channelization codes, then are summed to obtain the phase signal (I). The other 3 DPDCHs and the DPCCH are spread by means of channelization codes and are summed to form the quadrature signal (Q); in this case with a maximum PG value of 4, at most 4 different transmissions can be made in each branch with different OVSF codes. Finally, complex scrambling is adopted. Since the bit-rate per DPDCH is 960 kbit/s, the maximum aggregated bit-rate is 6*960 = 5760 kbit/s. Hence, depending on the coding scheme with rate 1/2 or rate 1/3, it is possible to achieve 1920 or even 2880 kbit/s information bit-rate values, with different bit error rate performance.

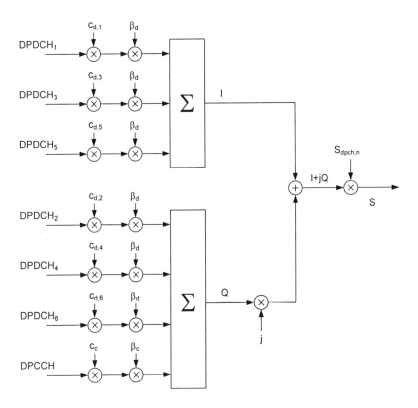

Fig. 44: Spreading process with 6 DPDCHs and one DPCCH.

Referring to Fig. 44, channelization and scrambling processes can be detailed as follows:

- The OVSF channelization code transforms every bit into a number of chips, thus increasing the bandwidth of the signal.

  - DPCCH with channelization code c_c = $C_{ch,256,0}$ corresponding to a bit-rate of 15 kbit/s,

  - Only one DPDCH (as in Fig. 43) with channelization codes $c_{d,1} = C_{ch,PG,j}$ $(j = PG/4)$ corresponding to a bit-rate of $3840/PG$ kbit/s $\in$ [15, 30, ..., 960] kbit/s for PG = 256, 128, ..., 4.

  - Or more DPDCHs (as in Fig. 44) with channelization code $c_{d,n} = C_{ch,4,j}$ and $j = 1$ for $n \in \{1, 2\}$, $j = 3$ for $n \in \{3, 4\}$, and $j = 2$ for $n \in \{5, 6\}$, because each couple of DPDCH is divided between I and Q channels.

After channelization, the spread signals are weighted by gain factors β_d for DPDCHs and β_c for DPCCH that range from 0 to 1.

- Scrambling code can be either short or long. Short codes are used to ease the implementation of advanced multi-user receiver techniques; otherwise, long spreading codes can be used. Short codes are S(2) codes of period 256 chips; long codes are Gold sequences of length 2^{25} - 1 chips that are truncated to obtain a period of 10 ms (i.e., 38400 chips at 3.84 Mchip/s). There are 2^{24} long and 2^{24} short uplink scrambling codes (no code reuse is needed among cells).

The scrambling process is in the complex domain. Let us now refer to a simple case with a single DPDCH, as shown in Fig. 45. Two consecutive bits of a channel (either DPDCH or DPCCH) are first serial-to-parallel converted; then, a complex scrambling process is performed with the circuit in Fig. 45.

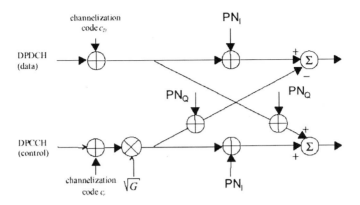

Fig. 45: I/Q complex scrambling scheme.

Finally, the complex-valued chip sequence generated by the spreading process is QPSK modulated by using a square-root raised cosine impulse shaping filter (Fig. 46).

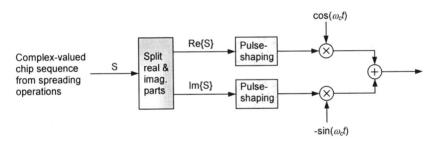

Fig. 46: Uplink QPSK modulation.

Since the spreading bandwidth is fixed irrespective of the user bit-rate, a higher transmission bit-rate on a single DPDCH is obtained by reducing PG and increasing the transmission power to maintain a given BER value (QoS requirement). This situation is depicted in Fig. 47, where different rates are supported with different power levels (= heights).

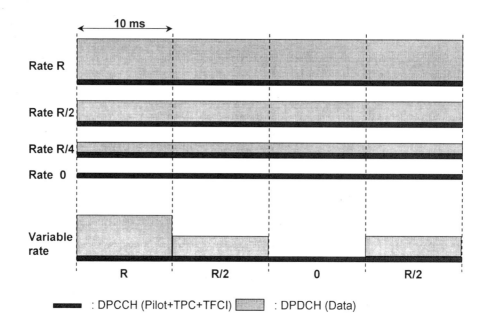

Fig. 47: Uplink variable-bit-rate transmissions.

Let us now focus on downlink transmissions. There is one downlink dedicated physical channel, one shared and several common control channels (some of them are listed below):

- *Downlink dedicated Physical CHannel* (DPCH)

- *physical Downlink Shared CHannel* (DSCH)

- *primary and secondary Common PIlot CHannels* (CPICH)

- *primary and secondary Common Control Physical CHannels* (CCPCH)

- *Synchronization CHannel* (SCH)

- *Acquisition Indicator CHannel* (AICH).

Referring to DPCH, Fig. 48 shows the frame structure [63], where DPDCH and DPCCH are time-multiplexed. DPCCH conveys control information generated at layer 1: known pilot bits, TPC bits for power-control commands and an optional TFCI field. DPCH can support variable bit-rate services when TFCI is transmitted or a fixed rate service when TFCI is not transmitted. The network determines if a

TFCI must be sent. DPDCH and DPCCH have the same power and the same processing gain value. In downlink, the processing gain value cannot vary on a frame basis (except for DSCH). When the total bit-rate to be transmitted on DPDCH exceeds the maximum bit-rate for a downlink physical channel, multi-code transmission is employed: several parallel downlink DPDCHs are sent using the same PG value. In this case, layer 1 control information is sent only on the first downlink DPCH.

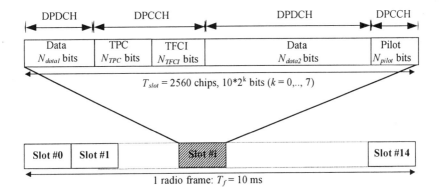

Fig. 48: Frame structure for downlink DPCH [63].

Downlink transmissions use the same OVSF channelization code tree adopted in uplink. Fig. 49 illustrates the spreading operation for all downlink physical channels except SCH (i.e., for AICH, DPCH, DSCH, CPICH and CCPCH); we can for instance refer to the DPCH case of Fig. 48.

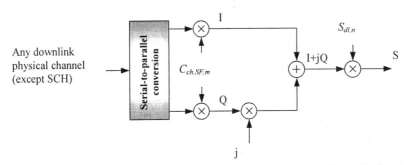

Fig. 49: Spreading process for a downlink channel (for all physical channels, except SCH); many of these signals are summed at the Node B to transmit to all the simultaneous users of a cell.

The non-spread physical channel consists of a sequence of real-valued *symbols*. For all the channels, except AICH, the symbols can take the three values +1, −1, and 0, where 0 indicates discontinuous transmissions. For AICH, the symbol values depend on the exact combination of acquisition indicators to be transmitted.

Referring to Fig. 49, the spreading process for a DPCH (i.e., DPDCH + DPCCH) is as follows: each pair of two consecutive symbols is first serial-to-parallel converted and mapped to I and Q branches (even and odd numbered symbols are mapped respectively to I and Q). Hence, the spreading process occurs with the same OVSF code for I and Q branches. The symbols that are conveyed on I and Q branches have a doubled duration with respect to the original one on DPCH. Hence, at a parity of spreading level, the downlink bit-rate conveyed by a DPDCH, $R_{down}(PG)$, doubles with respect to uplink (neglecting the few bits due to DPCCH): $R_{down}(PG) = 2*R_{up}(PG) = 7680/PG$ kbit/s, for PG = {4, 8, 16, ..., 512 for PG = 256. Hence, on a slot of duration 10/15 ms, we have a number of bits as a function of PG as $(7680/PG)*(10/15)$, corresponding to the following set of values: $\{10*2^k\}$ bits, with $k \in \{0, 1, ..., 7\}$. Of course the information bit-rate in downlink is reduced with respect to $R_{down}(PG)$ due to coding.

If a single DPDCH is not enough, three parallel DPDCHs can be transmitted (only one DPDCH is time-multiplexed with the DPCCH). Each DPDCH is serial-to-parallel converted and spread by one OVSF code. Them, all the I channels are summed; the same occurs for the Q channels.

Hence, the same spreading process occurs both in the cases of single and multiple DPDCH transmissions. In particular, I and Q branches are spread to the chip-rate by the same real-valued channelization code $C_{ch,PG,m}$. The sequences of real-valued chips on the I and Q branch are then treated as a single complex-valued sequence of chips. This sequence of chips is scrambled by a complex-valued scrambling code $S_{dl,n}$. Gold codes of length 2^{18} - 1 chips are used, but they are truncated to form a cycle of 10 ms (i.e., 38400 chips). To form a complex-valued code, the same truncated code is used with different time shifts on I and Q channels.

It is possible to generate $2^{18} + 1$ scrambling codes, but only 8192 of them are used. Each cell has allocated one scrambling code. In order to reduce the cell search time, scrambling codes are divided into 512 sets, each formed by a primary scrambling code and 15 secondary scrambling codes. Hence, each cell has one primary scrambling code. Secondary scrambling codes are used when one set of orthogonal channelization codes is not enough (e.g., when adaptive antennas are used in downlink).

The complex scrambled signal is QPSK modulated by adopting a suitable impulse shaping filer, $p(f)$. The process to generate the WCDMA downlink signal is described in Fig. 50.

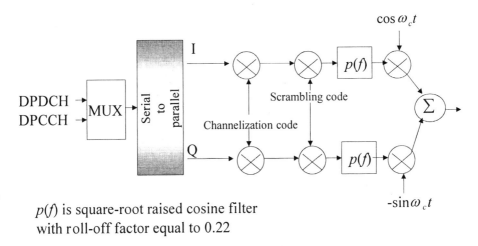

$p(f)$ is square-root raised cosine filter
with roll-off factor equal to 0.22

Fig. 50: Downlink QPSK modulation.

The resulting situation for downlink transmissions can be depicted as shown in Fig. 51.

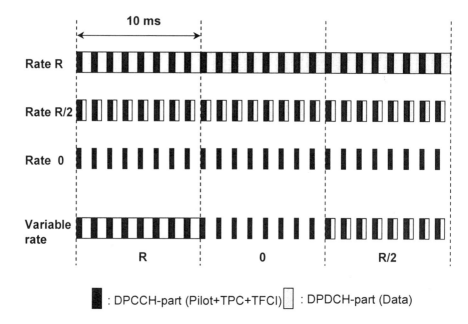

Fig. 51: Downlink variable-bit-rate transmissions with a fixed processing gain value.

3.4.2 Mapping of transport channels onto physical channels

The mapping of transport to physical channels is described in the Table

Transport Channels   **Physical Channels**

DCH ————————— Dedicated Physical Data Channel (DPDCH)

                Dedicated Physical Control Channel (DPCCH)

RACH———————— Physical Random Access Channel (PRACH)

CPCH ———————— Physical Common Packet Channel (PCPCH)

                Common Pilot Channel (CPICH)

BCH ————————— Primary Common Control Physical Channel (P-CCPCH)

FACH——————————— Secondary Common Control Physical Channel (S-CCPCH)

PCH ——————

                Synchronisation Channel (SCH)

DSCH———————— Physical Downlink Shared Channel (PDSCH)

                Acquisition Indicator Channel (AICH)

                Access Preamble Acquisition Indicator Channel (AP-AICH)

                Paging Indicator Channel (PICH)

                CPCH Status Indicator Channel (CSICH)

                Collision-Detection/Channel-Assignment Indicator

                Channel (CD/CA-ICH)

Table 11: Mapping of transport onto physical channels.

3.4.3   UTRA-TDD physical layer characteristics

In this air interface there is a TDMA frame of 10 ms with 15 slots (0.665 ms), as shown in Fig. 52 [55],[58],[67]-[70]. On each slot up to 16 parallel transmissions can be performed with different OVSF codes. UTRA-TDD has been completely standardized starting from UMTS Release 4 (March 2001) by harmonizing it with the proposed Chinese *Time-Division Synchronous CDMA* (TD-SCDMA) air interface. A major advantage of a TDD system is that higher overall data rate on one link can be temporarily satisfied by taking part of slots of the other link (if available). Hence, UTRA-TDD is well suited for asymmetric transmissions, as it occurs in streaming, interactive and background traffics.

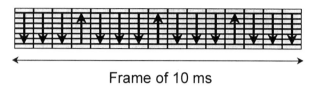

Frame of 10 ms

Fig. 52: UTRA-TDD air interface.

The chip rate of 3.84 Mchip/s for UTRA-TDD has been fully integrated with the chip rate of 1.28 Mchip/s, the value adopted by TD-SCDMA, a 3G standard that uses a 5 ms frame length with 7 traffic slots/frame.

A user may transmit using different OVSF codes on the same slot and/or on different slots. The transmission is made with QPSK modulation for both uplink and downlink. Joint detection can be used to reduce the multiple access interference due to transmissions of different users on the same slot. This is made possible by the packet structure shown in Fig. 53. A packet is transmitted on a given slot-code resource.

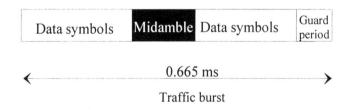

0.665 ms

Traffic burst

Fig. 53: UTRA-TDD burst (different burst types are available; see Part II of this book).

3.5   Voice service in UMTS

In Release '99, the user speech is digitally sampled by the mobile user equipment, and then coded for transmissions. The default speech coding (supported by UEs and the UTRAN) is *Adaptive Multi-Rate* (AMR). The AMR coder supports eight source rates ranging from 4.75 kbit/s to 12.2 kbit/s, and is rate-controlled so that it can rapidly switch between them at any point in the call. The bit-rates are selected depending on the required QoS level for speech and the quality of the transport provided by the network (primarily that of the radio link). The AMR coder also supports a low-rate background noise-encoding mode to reduce the bit-rate during silence periods, thus reducing bandwidth and battery usage in the user equipment. In addition to AMR, other speech coding may be optionally selected, such as *Enhanced Full Rate* (EFR) or *Full Rate* (FR), as specified for GSM. Within the core network, the speech coding at 64 kbit/s or 56 kbit/s (ITU-T Recommendation G.711) is generally used as in the PSTN and in GSM core networks. Transcoding from AMR (or other speech coding) to G.711 is performed in the MSC.

3.6   New service concepts supported by UMTS

Defined by 3GPP as a standard for third-generation implementation, the *Multimedia Messaging Service* (MMS) is an evolution of the textual SMS provided by GSM networks. There are virtually no limits to the contents of an MMS transmission: formatted text, graphs, data, animations, audio clips, voice transmissions and video sequences. Sending digital postcards and PowerPoint-style presentations is expected to become the killer application for MMS.

The success and the diffusion of novel services in 3G systems relies on the *Virtual Home Environment* (VHE) concept [80], which defines the framework for how applications can be developed and implemented by service providers. VHE enables the user personal service environment to be portable across network boundaries and between terminals. VHE ensures that users are consistently presented with the same personalized features, user interface and services in whatever network, with whatever terminal and wherever the user may be located. Therefore, the user always feels that he is on his home network even when roaming across network boundaries. Full security will be provided transparently across a mix of access and core networks.

3GPP defines the *Home Environment* (HE) as the environment that is responsible for the overall provision of services to users. The HE is within the domain of the user HPLMN. 3GPP proposes four methods by which service providers are able to provide VHE services and applications to their subscribers; they are described below.

- *Customised Applications for Mobile network Enhanced Logic feature* (CAMEL), CAMEL provides the mechanism to support *Intelligent Network* (IN) type services in the mobile network environment consistently and independently of the serving 3GPP network.

- *Mobile station Execution Environment* (MExE), allows service providers to develop applications that can be accessed through different types of terminals, regardless of the terminal platform. These applications that reside on the terminal can then interact with a MExE server for service provision. This approach is particularly suitable for terminal-centric services. Other details on MExE are given at the end of Chapter 9 in Part II of this book.

- *UMTS SIM Application Toolkit* (USAT) that provides VHE by allowing applications residing in the SIM card to interact with applications or services located on a server in the network.

- *Open Service Architecture* (OSA), developed for server-centric applications and services, allows to access network features and capabilities (e.g., call control) through a set of standardized interfaces.

Users may choose to adapt VHE services to different QoS requirements depending on the network, thus creating a user profile mapping "services to network QoS levels". For example, the service should be able to request higher QoS from the network provider when the network becomes less congested. 3GPP has proposed the use of a MExE QoS manager component residing in a MEXE compliant terminal to support these functionalities.

Finally, *LoCation Services* (LCS) are another important characteristic of future mobile communication systems. In fact, LCS provide information for users on the move related to their positions. Interesting LCS applications can be: emergency call services, tourist information services, pre-trip and on-trip information services, car theft recovery and so on. Both 2G and 3G standards have specified their LCS techniques that can be broadly divided into two categories:

- *Network-based LCS* (i.e., all LCS measurements and elaboration is performed by the network);

- *Mobile-assisted and/or based LCS* (i.e., either the mobile handset performs measurements and reports them to the network or the mobile handset actually obtains its position and communicates it to the network).

The following localization techniques have been standardized:

- *2G systems (GSM, GPRS):*
  - ➤ *Timing Advance* (TA)
  - ➤ *Enhanced Observed Time Difference* (E-OTD)
  - ➤ *Global Positioning System* (GPS)

- *3G systems (UMTS Releases '99, 4 and 5)*
  - ➤ Cell-ID based positioning system
  - ➤ *Observed Time Difference Of Arrival with network adjustable Idle Periods DL* (OTDOA-IPDL)
  - ➤ GPS based positioning system (GPS and Assisted-GPS).

3.7   UMTS releases differences

A roadmap was initially proposed for the development of UMTS in subsequent phases:

- *UMTS phase 1*, with the implementation of a new access network (UTRAN) connected to an evolved GSM – GPRS core network;

- *UMTS phase 2*, with the introduction of a UMTS core network and the complete deployment of system architecture and services.

Subsequently, 3GPP has organized the evolution of 3G systems according to Releases. In particular, we will describe below the main characteristics of "Release '99" (frozen in December 1999), "Release 4" (frozen in March 2001) and "Release 5" (frozen in March 2002).

3.7.1   Release '99

Main features:
- Creation of a new *Radio Access Network* (RAN) with FDD mode of operation

- Interoperability with GSM-GPRS networks.

Services and system aspects:
- Services as available with GSM
- MMS
- OSA, basic version
- LCS.

Applications supported by terminals:
- SMS as in GSM
- MMS
- MExE R99
- Multi-mode UEs

Core network protocols:
- CAMEL Phase 2 and 3
- Basic UMTS security issues
- ASCI (*Advanced Speech Call Items*) call forwarding enhancements
- GPRS elements
- GTP (*GPRS Tunneling Protocol*) enhancement
- Handover
- GSM-UMTS interworking
- Multi-call
- Circuit-switched bearers in UMTS
- OSA (basic version).

The first 3G network installed in Japan (FOMA system by NTT DoCoMo) is based on Release '99 specifications.

3.7.2   Release 4

Services and system aspects:
- *Transcoder-Free Operation* (TrFO)
- Tandem Free aspects for 3G and between 2G and 3G systems

- VHE and OSA evolution
- Full support of LCS in circuit-switched and packet-switched domains.

3G radio access:
- New TDD mode at 1.28 Mchip/s
- Evolution of UTRAN transport for IP support
- Various RAN improvements, such as robust header compression according to IETF RFC 3095.

Terminals supported applications:
- MExE - Rel4
- MMS - Rel4
- USAT interpreter protocol (partly Release 5)
- USIM related features from CPHS (*Common PCN Handset Specifications*)
- Logical channels for USIM.

Core network protocols:
- Evolution of transport in the CN
- Non-Transparent Real-Time Facsimile
- Enable bearer-independent CS architecture
- ASCI enhancements for Rel-4
- *Operator Determined Barring* (ODB) for packet-oriented services.

3.7.3   Release 5

The UMTS core network is divided into three domains:
- *Circuit Switched* (CS) domain,
- *Packet Switched* (PS) domain,
- *Internet protocol Multimedia* (IM) domain.

CS domain is a continuation of the GSM standard network infrastructure. Whereas, PS domain is an enhancement of the GPRS infrastructure mainly concerning QoS classes (conversational, streaming, interactive, background). IM domain is an all-IP network infrastructure for providing multimedia applications to mobile users:

- IM bridges the gaps between mobile networks and IP-based telephony/streaming, supports call control, session control and subscriber management,
- IM integrates GSM/UMTS SS7 and TDM protocols with SCTP, H.323, SIP and MPLS.

The IM architecture consists of IP capable terminals, gateways (signaling and media conversions) and call agents (call control, session management).

Release 5 is an all-IP system characterized as follows.

Services and system aspects:
- Development/selection of a multi-rate wideband speech codec with extended acoustic bandwidth (50 Hz - 7 kHz) for the support of wideband speech telephony in multiple radio environments
- Provisioning of IP-based multimedia services
- Support of push services
- Enhancements to security, VHE, OSA, global text telephony and LCS.

3G radio access:
- Intra-domain connection of RAN nodes to multiple core network nodes
- *High Speed Downlink Packet Access*, HSDPA, (maximum downlink bit-rate up to 10 Mbit/s)
- RAN improvements to enable efficient IP-based multimedia services in UMTS
- Separation of resource reservation and radio link activation (offers benefits to high bit-rate users).

Terminals supported protocols:
- MExE - Rel5
- MMS - Rel5

- Enhancements to USIM toolkit secure messaging.

Core network protocols:

- Provisioning of IP-based multimedia services (SIP call control protocol)
- PS emergency call enhancements
- CAMEL Phase 4
- Intra-domain connection of RAN nodes to multiple core network nodes
- Reliable end-to-end QoS for the PS domain.

Chapter 4: Satellite communications

In October 1945 a RAF electronics officer and member of the British Interplanetary Society, Arthur C. Clarke, wrote an article in the *Wireless World* journal entitled "Extra Terrestrial Relays – Can Rocket Stations Give Worldwide Coverage?" that described the use of *manned* satellites in orbits at 35,800 km altitude, thus having synchronous motion with respect to a point on the earth (i.e., 24-hour orbital period). Such characteristic suggested him the possible use of these *GEOstationary* (GEO) satellites to broadcast television signals on a wide part of the earth. In fact, three GEO satellites would be enough to cover all the earth except Polar Regions.

Satellites are in elliptical orbits around the earth according to the Kepler's laws.

Clarke's article apparently had small effect. Perhaps the first person who carefully evaluated the various technical options as well as economical aspects in satellite communications was John R. Pierce of AT&T's Bell Telephone Laboratories. In 1955, he described in an article the utility of a communication "mirror" in space, a medium-orbit "repeater" and a 24-hour-orbit "repeater". He proved the usefulness of satellites as compared to transatlantic telephone cables.

After the launch of Sputnik I in 1957, many considered the benefits, and the profits associated with satellite communications. The first experimental telephone and TV transmissions occurred with *Low Earth Orbit* (LEO) and *Medium Earth Orbit* (MEO) satellites in years 1962-1964:

- TELSTAR I (77 kg, 940-5640 km)
- RELAY I (78 kg, 1320-7430 km)
- TELSTAR II (79 kg, 970-10800 km).

From 1963, GEO satellites experiments were put in place to prove the effectiveness of satellite communications:

- SYNCOM II (39 kg, almost GEO)
- SYNCOM III (66 kg).

In August 1964, INTELSAT was formed as an international organization for satellite communications.

Since then satellites have been used to provide communication over long distances (very popular before the advent of fiber) and over wide coverage areas (primarily for broadcasting TV signals). In particular, we can recall here the following important milestones:

- 1965: INTELSAT I ("Early Bird") begun GEO intercontinental telecommunication services
- 1972: FIXED CONTINENTAL SERVICES begun US regional communication services
- 1982: INMARSAT GLOBAL SYSTEMS was fully operational for mobile maritime services
- 1988: OMNITRACS, the first land mobile satellite system started to provide in North America land mobile satellite messaging and localization services
- 1991: ITALSAT (Italy) was operational, the first satellite with on board processing and multi-beam coverage.

The INTELSAT system employs a network of geostationary satellites located in orbital positions over the Atlantic, Indian, and Pacific Oceans. INTELSAT is composed of over 130 member nations.

INTELSAT I total capacity was only 240 two-way voice circuits or one television (TV) circuit. The operational frequency was in C-band with a transmit power of 6 Watts. Current INTELSAT systems are based on four generations of bent-pipe spacecraft including INTELSAT V/V-A, VI, VII/VII-A, and VIII. Every new generation of INTELSAT satellites has been designed to accommodate increasing capacity. INTELSAT VII has a total capacity is 18,000 two-way voice circuits plus 3 TV circuits.

INTELSAT provides a wide range of international services including telephone, data transfer, facsimile, television broadcasting, and teleconferencing. INTELSAT leases space segment capacity with increments of 9, 18, 36, 54, or 72 MHz. Access to the INTELSAT satellites is through a wide array of terminal types with antenna sizes ranging from 0.5 m to 30 m and operating frequencies in both the C-band (6 GHz uplink; 4 GHz downlink) and Ku-band (14 GHz uplink; 11 or 12 GHz downlink).

4.1 Basic considerations on satellite communications

Satellites typically require *Line-Of-Sight* (LOS) communication and the signal is too weak to penetrate buildings. If the service is to be used indoors, some other form of communication is needed to relay the signal from the LOS antenna.

Although initially only used as relay stations (i.e., bent-pipe satellites), satellites now can regenerate the signal (i.e., they demodulate, correct and retransmit the signal) and can have intelligent switching capabilities on board.

Satellites are typically bandwidth and power limited. Hence, their resources must be efficiently used. Thus, modulation methods, coding schemes and antenna systems need to be suitably designed to meet adequate QoS levels with the link budget.

Each satellite has a *user link* with remote terminals / stations (end-users) and a *feeder link / control link* with an earth control station. As for the user link, typically a multi-beam antenna is used on the satellite in order to divide the covered territory in smaller areas. These cells (irradiated by beams) are used to concentrate the transmitted energy and also to tailor the satellite service area on the earth. The capacity of the network can be increased through frequency reuse among the different beams.

4.1.1 Satellite orbit types

Opposed to *Wireless Local Area Networks* (WLAN) that offer high bit-rates over limited geographic distances, satellite networks provide wide coverage areas with low bit-rates. Different satellite orbital options are available depending on the altitude of satellites [81]. Of course, the lower the altitude the higher the number of satellites to cover all the earth. Satellite coverage is organized in planes, as shown in Fig. 54.

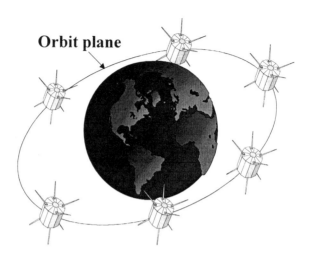

Fig. 54: Satellites on a given orbit plane (covering a belt on the earth).

With *satellite-fixed cell* systems, beams maintain a constant geometry with respect to the spacecraft and the cells on the ground move along with the satellite. With *earth-fixed cell* systems, the antenna beams are steered so as to point towards a given cell on the earth as much as possible during the satellite visibility time. The *satellite-fixed cell* technique is for instance employed by the operational LEO system named Globalstar [82]. The *earth-fixed cell* technique will be employed by future LEO multimedia satellite systems such as Teledesic [83] and Skybridge [84].

Satellite motion characteristics are tabulated in the satellite ephemerides.

The main satellite orbital options are detailed below.

- *GEostationary Orbit* (GEO)

  A GEO satellite has a circular equatorial orbit with a 24-hour period at an altitude of about 35,800 km. As a result, the satellite remains stationary over the same point on the earth. Depending upon the antenna design, a single satellite in a GEO orbit can illuminate a spot beam that covers approximately 32% of the earth surface, assuming a minimum ground antenna elevation angle of 10°.

Hence, three GEO satellites are sufficient to cover all the earth, except Polar Regions (additional satellites serve as backup). A significant drawback for GEO transmissions is the high signal propagation delay. Such delay is not compatible with real-time transmissions of data, but it is compatible with non-interactive transmissions, such as file transfers or video broadcast.

- *Low Earth Orbit* (LEO)

These satellites are in orbits at 500-2000 km altitude (below the inner Val Allen radiation belt that is dangerous for electronic equipment and above the higher layers of the atmosphere that would cause the orbit to be unstable due to drag), thus reducing the propagation delay problem. However, a LEO satellite does not remain stationary relatively to surface locations. Hence, many satellites on the same orbit plane must alternate in covering a given area (see Fig. 55).

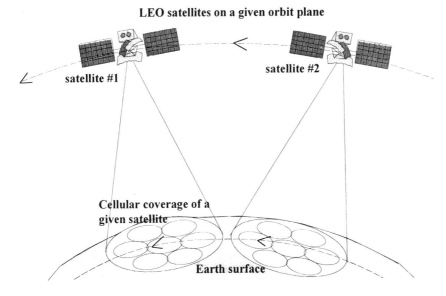

Fig. 55: LEO coverage for a given area: many satellites alternate to cover this area (satellite-fixed cell case).

Note that orbit planes have a periodic motion due to the earth rotation around its axis.

Sophisticated techniques must be implemented in earth control stations to track LEO satellites; inter-beam and inter-satellite handoff procedures must be implemented (beam changes are frequent during a communication). The received signal can be affected by significant Doppler frequency shifts due to the relative radial speed, so that some countermeasures need to be adopted in digital communication systems[4].

The lower the satellite orbit, the faster it moves, and the smaller the covered area on the earth. The satellite orbital speed can be found by equaling the earth gravitational attraction to the centrifuge force:

$$\gamma \frac{m_T m_s}{\left(R_T + H\right)^2} = m_s \frac{V_{orb}^2}{R_T + H} \quad \Rightarrow \quad V_{orb} = \sqrt{\frac{\gamma m_T}{R_T + H}} \qquad (14)$$

where m_s is the satellite mass, m_T is the earth mass, γ is the gravitational constant, R_T is the mean earth radius, H is the satellite constellation altitude, V_{orb} is the satellite orbital speed.

The satellite angular velocity is $\omega_s = V_{orb}/(R_T + H)$. Hence, the satellite orbital period, P, is equal to $2\pi/\omega_s$. Moreover, the satellite ground-track speed, V_{trk}, is obtained by using the proportionality suggested by Fig. 56.

$$\frac{V_{orb}}{V_{trk}} = \frac{R_T + H}{R_T} \quad \Rightarrow \quad V_{trk} = \frac{R_T}{R_T + H} V_{orb} \ . \qquad (15)$$

The behavior of V_{trk} as a function of the satellite constellation altitude, H, has been shown in Fig. 57; we may note that V_{trk} is quite high, varying from 19,000 to about 26,000 km/h depending on the satellite constellation altitude.

LEO systems require many satellites (40 to 70) that orbit in a carefully controlled pattern (= constellation), where satellites of different planes rotate with suitable *phases*. The high cost of a fleet of satellites and the complexity of the control system greatly increase the cost of a LEO satellite constellation.

---

[4] The frequency of the received signal is increased (reduced) of $f_D = v/c$ (where v is the radial speed and c is the light speed) with respect to the frequency of the transmitted signal.

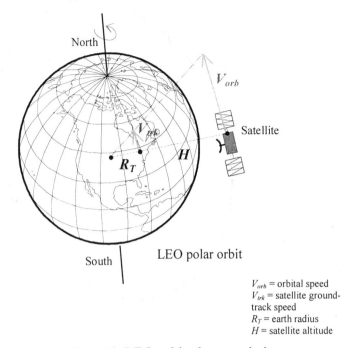

Fig. 56: LEO orbit characteristics.

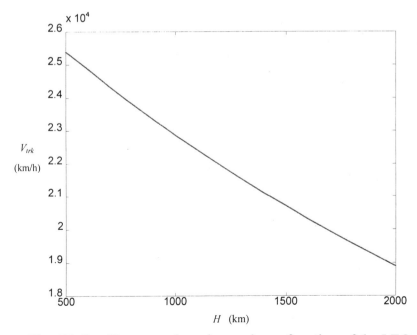

Fig. 57: Satellite ground-track speed as a function of the LEO altitude.

-   *Medium Earth Orbit* (MEO)

    These satellites are at intermediate altitudes around 10,000 km. From 10 to 15 MEO satellites are needed to serve all the earth.

-   *Highly Elliptical Orbit* (HEO). This solution is well suited for regional coverage.

Finally, hybrid orbital solutions are also possible, that is satellite systems including some GEO and some non-GEO satellites in order to get the best of both cases (e.g., Spaceway [85]).

LEO and GEO systems can be compared as detailed in Table 12.

GEO	LEO
Advantages - Simplicity of the satellite configuration and constant satellite position - Well-known maintenance and control aspects - Doppler effects of minor impact - Extremely wide spot-beam footprint - Numerous launch opportunities - Few large, complex, high-power satellites with long life cycle *Disadvantages* - Considerably high signal delay (about 125 ms for each link) - More expensive class of launchers - Low elevation angle on ground for latitudes grater than 50° - High signal attenuation	*Advantages* - Small propagation delays - Reduction of the power requirements on-board and on the ground - A lower class of launchers is needed *Disadvantages* - Many flight units are needed in order to provide coverage requirements. - Complex orbital design and orbital maintenance - Frequent handoff procedures between beams of the same satellite or between beams of adjacent satellites - Spacecraft tracking can be necessary for the user terminal - Significant Doppler effects - Many ground stations or inter-satellite links (expensive system).

Table 12: LEO and GEO advantages and disadvantages.

Fig. 58 gives a pictorial view of different satellite orbital options with relevant planned or operation satellite systems for mobile personal communications.

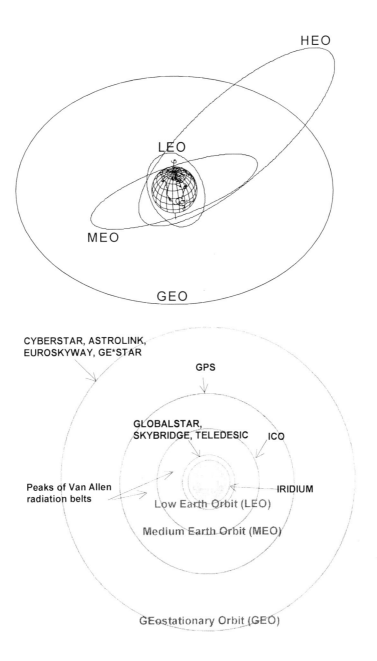

Fig. 58: Satellite orbital options and some satellite systems.

The *Round Trip propagation Delay* (RTD) depends on the satellite altitude and the minimum elevation angle (mask angle) accepted on the earth for the link to the satellite. High RTD values prevent an immediate feedback to users (this may have impact on access protocol performance). Of course, the higher the satellite constellation altitude the greater the delay. A given RTD value can be obtained with several combinations of minimum elevation angle and satellite constellation altitude (Fig. 59).

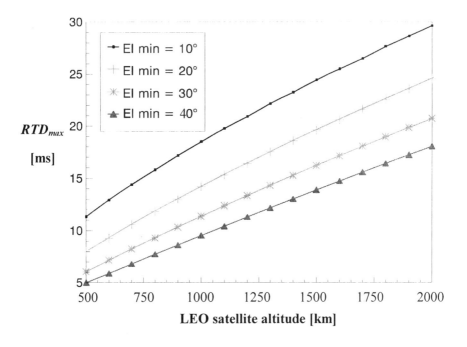

Fig. 59: RTD values for LEO systems.

RTD values in GEO systems are about 250 ms (this delay doubles if we consider a transparent GEO satellite, so that all the control information must be sent from the GEO satellite to an earth station) and prevent immediate feedback to terminals. Moreover, high RTD values cause a noticeable and annoying echo in telephone calls.

LEO satellites are at low altitudes, thus allowing low RTD values.

Medium and high RTD values create some problems to interactive and real-time services and to some data protocols such as TCP and HDLC. Such protocols need to be modified for the satellite environment (timer values and window sizes).

4.1.2   Frequency bands and signal attenuation

The typical frequency bands used in satellite communications are as follows.

- C band (4-8 GHz) that is already congested.
- Ku band (10-18 GHz) used by the majority of satellite digital broadcast systems as well as current Internet access systems (e.g., DirectPC[5] and Starband[6] [86],[87]).
- Ka band (18-31 GHz) that offers higher bandwidths with smaller antennas, but presents significant attenuation.

Of course the higher the frequency band, the higher the attenuation according to the free-space path loss, L_{free}:

$$L_{free} = \left(\frac{4\pi D}{\lambda} \right)^2, \quad where \ \lambda f = c \qquad (16)$$

being D the earth-to-satellite distance, λ the transmitted signal wavelength, f the transmission carrier frequency and c the light speed.

Additional attenuation is due to the presence of the atmosphere. Attenuation peaks are at 22.3 (within Ka band) and at 60 GHz (within V band) respectively due to water vapor and molecular oxygen.

---

[5] DirectPC system is based on an asymmetric access to the Internet: uplink (for user requests) is made through the fixed terrestrial network (typically, 14-56 kbit/s), whereas the downlink high-capacity channel (downloading) is provided through the satellite (typically, 28 Mbit/s effective bandwidth). DirectPC system requires a 18-inch diameter antenna and a set-top box.

[6] Starband is a two-way always-on high-speed Internet access service via satellite implemented in the U.S.A. Uplink and downlink transmissions are via satellite.

4.1.3   Satellite network telecommunication architectures

Two different types of satellites can be considered:

- *Bent-pipes satellites* (i.e., satellites act as repeaters). Signal is amplified and retransmitted, but there is no improvement in the C/N ratio, since there is no demodulation, decoding or other type of processing.

- *Satellites with On-Board-Processing*: satellites demodulate and decode the received signal, thus achieving signal recovery before transmitting it. Since at some point base-band signals are available, other activities are also possible, such as routing and switching. Hence, these satellites allow the use of *Inter-Satellite Links* (ISLs) with other satellites of the same constellation, thus permitting the signal routing in the sky.

Correspondingly, two different satellite system architectures have been conceived, as shown in Figs. 60 and 61.

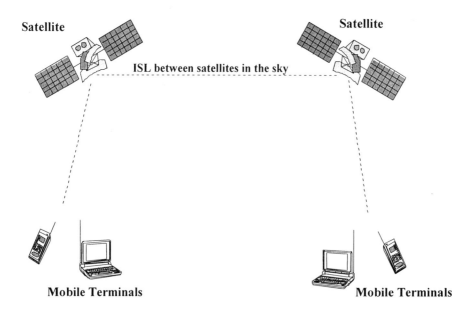

Fig. 60: Satellite system with ISL (i.e., routing in the sky, as proposed by the Teledesic system [83]). Intra-plane and inter-plane ISL may be supported.

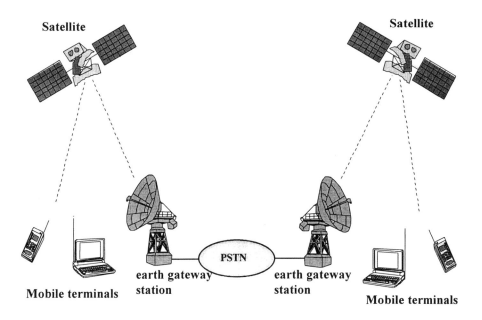

Fig. 61: Satellite system without ISL (the signal from the satellite is transponded to the earth gateway stations, as in the Globalstar system [82]).

In the case of a LEO satellite constellation with ISLs, the routing is quite complex due to the satellite position dynamics.

4.2   Different types of mobile satellite systems

Satellites are used for a variety of purposes including sensors and data collection (e.g., Landsat, ARGOS, Defense Satellite Program), weather (e.g., GOES, Defense Meteorological Satellite program), navigation and timing (e.g., GPS), weapons, reconnaissance, and communications (e.g., INTELSAT).

The interest is here on the adoption of satellite systems for mobile personal communications.

The role of mobile communication systems through satellites can be identified as follows [88],[89]:

- *Coverage completion and extension*: due to the large coverage area offered by a satellite beam, the mobile satellite system can be used to complete the coverage of terrestrial cellular networks. This is particular useful in order to provide services to spread-out populations. Moreover, aeronautical and maritime users may be solely dependent on the satellite component for the provision of mobile personal communications.

- *Dynamic traffic management*: the satellite resource can be used to off-load some of the traffic from the terrestrial cellular network.

- *Disaster-proof link availability*: satellite systems can provide a back-up service in cases of disasters and in any situations in which a terrestrial network is malfunctioning.

- *Global roaming*: satellite systems can provide users with global roaming capability.

4.2.1   Satellite UMTS

ETSI specifications foresee a satellite access to the UMTS network: *Satellite-UMTS* (S-UMTS). The satellite sub-system should realize the *UMTS Satellite Radio Access Network* (USRAN) connected to the UMTS core network via the Iu interface [90]. USRAN is expected to be implemented in 2005-2006 and should be able to support user bit-rates up to 144 kbit/s. This bit-rate is sufficient to provide multimedia services based on the H.320, H.323, H.324 and MPEG-4 standards [91].

As for the QoS and end-to-end delay requirements, a target BER equal to 10^{-3} and a maximum delay of 400 ms have been envisaged for speech services, while a BER of 10^{-6} have been retained for data applications. Different delay figures have been considered for each class of data service (e.g., few seconds for Internet access and few minutes for e-mail delivery).

It is expected that the S-UMTS system will be capable of supporting a range of existing and future applications. In fact, S-UMTS will not only complement the coverage of the *Terrestrial UMTS* (T-UMTS), but it will also extend its services. The typical operational environments for S-UMTS are therefore areas where the T-UMTS coverage would be either technically or economically not viable (see Table 13).

Operational environments	T-UMTS	S-UMTS
Maritime areas	No	Yes
Aeronautic	No	Yes
Rural Areas	No	Yes
Highways	Yes	Yes
Suburban areas	Yes	Yes
Urban areas	Yes	No
Indoor	Yes	No

Table 13: UMTS usage scenarios.

Satellite services (with the exception of low bit-rate services such as paging) will be mainly provided under LOS propagation conditions.

The table below highlights the six RTT (*Radio Transmission Technology*) proposals submitted to ITU as satellite component of IMT-2000 on June 1998.

Proposals based on WCDMA	Proposals based on hybrid air interfaces (CDMA/TDMA)	Proposals based on TDMA
SW-CDMA by ESA	SW-C/TDMA by ESA	ICO RTT by ICO Global Telecommunications
Sat-CDMA by TTA S. Korea	INX by Iridium Operating LLC	Horizons by Inmarsat

Table 14: Proposals for S-UMTS.

All these proposals foresee the support of data transmissions up to 144 kbit/s and can be considered as adaptations of the corresponding terrestrial air interfaces.

A WCDMA air interface has been standardized by ETSI for S-UMTS [92].

4.2.2 Future satellite system protocols for high-capacity transmissions

Satellite networks are foreseen to provide broadband services to geographically dispersed user groups. The need for broadband services call for the support of suitable protocols. It is anticipated that satellite multimedia systems will adopt the following protocols [93]:

-   TCP/IP,
-   DVB-S digital platform,
-   ATM.

Let us focus now on the Internet access through INTELSAT satellites. More than 80% of African countries use INTELSAT for their Internet traffic with Europe and North America. Internet transmissions via satellites are attractive due to:

-   Rapid and easy implementation
-   Network congestion bypass
-   High quality.

Internet traffic is asymmetric; the ratio of request traffic to return traffic is typically 1:10. Satellite links address Internet asymmetry in a cost-effective way by allowing asymmetric resource allocations.

Web content is cached in order to make a faster access to most common multimedia information on the Web. According to this technique, popular Web content is pushed closer to the end-user, thus allowing a faster access, improved bandwidth efficiency and transmission cost reduction.

Internet access for rural communications can be achieved through *Very Small Aperture Terminal* (VSAT) via INTELSAT satellites. VSAT is developed for fixed broadband access via satellite. Fig. 62 describes the architecture for the satellite access through VSAT for local ISPs. Original VSAT systems where designed to operate in the Ku band. Now Ka band is adopted. TDMA or a special combination of FDMA and TDMA, named *Multi-Frequency TDMA*[7] (MF-TDMA) [94] are used as access technologies for VSATs.

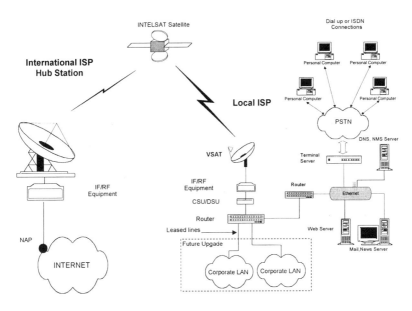

Fig. 62: Example for the use of VSAT: remote high-capacity access to the Internet.

4.3   Overview of proposed mobile satellite systems

This Section presents some satellite systems for personal communications [82]-[85],[95]-[96]. Table 15 surveys the major characteristics of planned and operational LEO systems. Some satellite systems are described in more details below.

---

[7] MF-TDMA air interface uses a frame with different slots. On each slot, a terminal can transmit its packet on a carrier to be chosen among a given number.

LEO System	Coverage	No. of satellites, altitude, cells, mask angle	Medium access scheme	Inter-sat. links	Freq. Bands
Operational Big LEO Systems					
Iridium	Global	66, 780 km, satellite-fixed cells, 9°	48 beams/satellite FDMA/TDMA	Yes Links at: 23.18-23.38 GHz	Uplink: 1616-1626.5 MHz Downlink: 1616-1626.5 MHz
Globalstar	±70°	48, 1414 km, satellite-fixed cells, 10°	16 beams/satellite FDMA/CDMA	none	Uplink: 1610-1626.5 MHz Downlink: 2483.5- 2500 MHz
Planned Big LEO systems					
Skybridge	±70°	80, 1469 km, earth-fixed cells, 30°	18 beams per satellite FDMA/ TDMA/ CDMA	none	Uplink: 12.75 to 14.5 GHz (Ku) Downlink: 10.7 to 12.75 GHz (Ku)
Teledesic	Global	288, 1375 km, earth-fixed cells, 40°	Uplink: FDMA/TDMA/ SDMA Downlink: ATDMA	1 Gbit/s	Uplink: 27.5-30 GHz (Ka) Downlink: 17.8-20.2 GHz (K)
Ellipso	Northern hemisph. and southern up to – 50°	2 elliptical inclined orbits 5 Borealis sats./plane 7846/520 km, 1 equatorial circular orbit, 6 Concordia sats., 8040km	61 beams/satellite FDMA/CDMA	none	Uplink: 1610-1626.5 MHz Downlink: 2483.5- 2500 MHz
Leoone (*Small LEO*)	±65°	48, 950 Km, 50° inclined orbits, 15°	random access and FDMA	none	Uplink: 148.00-150.05 MHz Downlink: 137.00-138.00 MHz, 400.15-401.00 MHz

Table 15: Main characteristics of LEO systems.

Globalstar

Globalstar system is able to provide communication services on all the earth, except Polar Regions (the covered area is within latitudes from 70° South and 70° North). Services were started in the third Quarter of

1999. Globalstar is composed of 48 satellites (6 satellites per plane) plus 4 spare spacecrafts on circular inclined orbits at 52°. Each satellite has 16 beams (in both uplink and downlink for the transmissions with users) and is at an altitude around 1414 km (i.e., LEO constellation). Satellites are built by Space Systems Loral and Alenia Aerospazio in Rome, Italy.

Fig. 63 shows the hypothetical coverage contour at 0° elevation angle for the Globalstar satellites on a generic orbit plane.

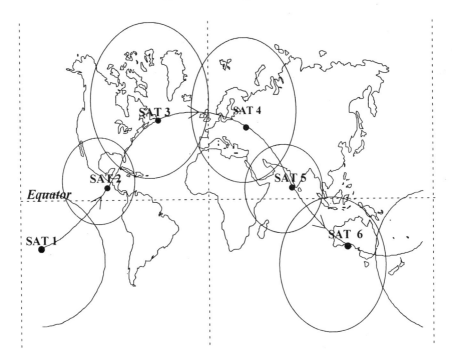

Fig. 63: Coverage areas for the satellites of a given Globalstar orbit plane (Mercator's projection view).

Each footprint moves rapidly across the earth surface. A satellite is visible only for about 15 minutes.

Globalstar air interface is based on a *synchronous Code Division Multiple Access*. The spacecraft employs links to control stations (C-band, 6875-7055 MHz) and handsets (L-band, 1610-1626.5MHz uplink; S-band, 2483.5-2500 MHz downlink).

The satellite constellation is designed to guarantee that at least two satellites can be seen by the users with an elevation angle greater than 15°; this characteristic allows to exploit some form of *satellite macro-diversity* that is important for the CDMA multiple access scheme adopted by the Globalstar system (i.e., the signals from/to two satellites are combined for an improved quality of communication). This technique also permits the implementation of a soft-handoff procedure.

Globalstar is a bent-pipe satellite and no ISL is adopted. Hence, each mobile terminal communicates via satellite directly to the nearest earth gateway station (gateways cover a radius of approximately 2000 km). The call is routed through the terrestrial telephone network (see Fig. 61). If another Globalstar mobile terminal must be reached, the call is routed to the closest gateway station and then relayed by the satellite to the mobile station. Therefore, a double satellite hop is needed. Propagation delays are anyway very low; for a mobile-to-mobile call they are about 72 ms.

Each satellite weighs 450 kg (dry mass of 350 kg) and is designed for a 7.5 year lifetime. The payload antennas are phased arrays mounted on the satellite.

The Globalstar system offers global, digital real-time voice, data and fax. Voice is encoded at a variable bit-rate (2.4, 4.8 or 9.6 kbit/s) depending on the background noise level. Data rates up to 9.6 kbit/s are available. Globalstar mobile phones are dual-mode with the possibility of accessing terrestrial cellular networks.

The Globalstar system supports phones that are either ANSI-41-based or GSM-based over the common *Globalstar Air Interface* (GAI). The Globalstar gateway provides the ANSI-41 capability and inter-works with an MSC for GSM support.

There are both vehicle mounted terminals and common hand-held phones. However, Globalstar also considers the deployment of fixed phone sites connected to the PSTN, a solution particularly attracting to provide communications in underdeveloped regions.

The Globalstar system is composed of three parts:

- ***Ground Operations Control Centers* (GOCCs)**

  A GOCC control the operation of satellites by gateway stations and coordinate with the *Satellite Operation Control Center* (SOCC). GOCCs plan the communication schedules for the gateways and control the allocation of satellite resources to each gateway.

- ***Satellite Operations Control Center* (SOCC)**

  SOCC manages the Globalstar satellite constellation. SOCC tracks satellites, controls their orbits, and provides telemetry and command services. Globalstar satellites continuously transmit spacecraft telemetry data and status reports. SOCC and GOCC communicate through the *Globalstar Data Network* (GDN).

- ***Globalstar Data Network* (GDN)**

  The GDN is the connective network that provides wide-area intercommunications facilities for the gateways, GOCCs and SOCC.

Spacecrafts are operated from a San Jose (CA) control center, and more than 60 further stations (including telemetry command unit gateways located in Aussaguel, France; Yeoju, South Korea; Dubbo, Australia; Bosque Allegre, Argentina; and Clifton, Texas) support the system.

The Globalstar system belongs to the first generation of satellite systems for mobile personal communications. Below a short survey is given about planned broadband satellite systems.

Skybridge

SkyBridge is a LEO satellite system designed to provide global access to interactive, multimedia communications, including Internet access and high-speed data communications.

The system is based on a constellation of 80 LEO satellites. Each satellite is in a circular orbit at an altitude of 1469 km above the earth. The constellation is divided into two symmetrical Walker sub-constellations (40 satellites each). The orbital inclination is 53°. A satellite has a 3000 km radius coverage divided into fixed spot beams of 350 km radius.

SkyBridge employs a combined code/time/frequency-division multiple access (CDMA/TDMA/FDMA) scheme. Spot beams, with frequency reuse in each beam, are employed to enhance capacity. Uplink operates at 12.75-14.5 GHz and downlink is at 10.7-12.75 GHz: the choice of the Ku-band is due to the availability of Ku-band technology. SkyBridge satellite design is based on a bent-pipe relay architecture; no ISL is considered. The architecture of the receiving sites may be adapted to the following configurations: individual reception, community reception (a SkyBridge terminal is shared between several subscribers) and professional configurations, where the SkyBridge terminal is connected to a LAN or a PBX. Each user is attached to only one SkyBridge gateway. The traffic to and from any of the SkyBridge users will always be concentrated on the gateway where the user is registered. SkyBridge gateway stations interface with terrestrial networks via ATM switches. The gateway handles interconnections with local servers and with terrestrial telecommunication networks.

Owing to the use of LEO satellites, SkyBridge propagation times are similar to those of landline broadband transmission systems (i.e., about 20 ms). This means that interactive multimedia traffics can be supported. Typical rates range from 16 kbit/s to as high as 60 Mbit/s. The majority of services are expected to be IP-based. SkyBridge will provide worldwide coverage starting with around 40 satellites. The total capacity for the completed constellation will be over 20 million simultaneously connected end-users.

Astrolink (2003)

The Astrolink satellite constellation contains nine GEO satellites operating in the Ka-band (uplink is 28.35-28.8 GHz and 29.25-30.0 GHz; downlink is 19.7-20.2 GHz). Astrolink satellites employ on board processing&switching. This system is designed to support high-speed multimedia communications. Data rates range from 16 kbit/s to 9.6 Mbit/s. 384 kbit/s are supported with 90 cm dish antennas that make Astrolink potentially suitable for large mobile platforms.

Cyberstar

Originally planned to use three GEO satellites in the Ka-band, now provides broadband data traffics with existing Telstar Ku-band satellites. The availability of a dedicated GEO constellation has been delayed. Cyberstar is designed to allow VSAT access with a bandwidth up to 45 Mbit/s. Cyberstar provides IP multicasting services to ISPs, large and small business organizations and multimedia content providers.

Spaceway (2002)

The Spaceway final configuration plans for 16 GEO Ka-band satellites and 20 MEO Ku-band satellites. This system is designed to support high-speed data, Internet access (i.e., next-generation DirectPC), and broadband multimedia information services. The Spaceway architecture is based on conventional bent-pipe relay satellites. It offers high QoS (BER lower than 10^{-10}) to users with terminals with antenna diameters as small as 0.66 m, at data rates starting from 16 kbit/s up to 6 Mbit/s. The Spaceway system is compatible with ATM, *Integrated Services Digital Network* (ISDN), frame relay and X.25 terrestrial standards.

Teledesic (2004)

The Teledesic constellation consists of 288 satellites in 12 planes (each with 24 satellites). Teledesic is a Ka-band system (uplink at 28.6-29.1 GHz and downlink at 18.8-19.3 GHz). It uses signals at 60 GHz for ISLs between adjacent satellites. Teledesic employs on board processing&switching. The system is designed to realize the "Internet in the sky". It offers high-quality voice, data, and multimedia information services. This system is designed for a BER lower than 10^{-10}. Multiple access is a combination of *Multi-Frequency TDMA* (MF-TDMA) on uplink and *asynchronous TDMA* (ATDMA) on downlink. The capacity of the network is planned to be 10 Gbit/s. Users will experience 2 Mbit/s on uplink and 64 Mbit/s on downlink. A minimum elevation angle of about 40° will permit that the Teledesic system reaches an availability of 99.9%.

WildBlue

WildBlue (formerly iSky and KaStar), is focused on providing broadband data services to North America. This is a Ka-band system able to support high-speed two-way Internet access and *Direct Broadcast Services* (DBS) to homes and offices via small aperture antennas. The initial constellation consists of 2 GEO satellites. The uplink frequency is at 19.2-20.0 GHz and the downlink at 29.0-30.0 GHz. Data rates up to 40 Mbit/s are considered with typical user bit-rates in the range of 1.5–5 Mbit/s.

High Altitude Platform Stations

Some designs for regional multimedia systems have replaced satellites with balloon-based telecommunication platforms (project ConSolar/Rorostar), *High Altitude Long Operational* aircraft (HALO) and *High Altitude Platform Stations* (HAPS). These platforms will be able to stay in the stratosphere at an altitude of about 21 km. A metropolitan area can be served within a 100 km radius by one beam of the geostationary satellite, by six-to-nine beams of a LEO satellite, and by as many as 700-to-1000 beams from a stratospheric platform.

One of the most significant projects is the HAPS system named SkyStation by Aerospatiale, Finmeccanica, Alenia Aerospazio, Thomson, Dornier or Comsat. An aircraft (157 m in length and 62 m in diameter) is equipped with solar batteries, which will feed a 1000-kg weight payload. Skystation platforms do not require a launch vehicle, they can move under their own power and they can be brought down to earth to be refurbished and re-deployed. The Skystation 47 GHz broadband service is based on 250 platforms worldwide, each operating independently and initially connected via ground-stations and existing public networks. Future platforms will be equipped with platform-to-platform links. Each aircraft has a coverage area with a diameter of 150 km. Up to 130 antennas (7 spots each) can be supported by an aircraft. 100 MHz spectrum is available in the 47 GHz band for both uplink and downlink. Each user will be provided with 2 Mbit/s in uplink and 10 Mbit/s in downlink. QPSK modulation will be adopted for the link to the user and 64QAM will be employed for the link with ground stations. FDMA/TDMA is the access scheme for uplink; TDM is employed in downlink. The services supported by Skystation are: mobile/portable telephony, video telephony, videoconference, high speed Internet access, video on demand. A 2 GHz payload is also considered for Skystation. One platform at an altitude of 21 km can provide 3G mobile communication services with the terrestrial WCDMA air interface in order to achieve umbrella coverage and to allow a rapid deployment of these new services.

Chapter 5: Mobile communications beyond 3G

Every 10 years we have assisted to a new generation of mobile communication technologies (see Fig. 64): from first-generation (1G) analog systems to second-generation (2G) digital networks to third-generation (3G) cellular systems. One of the reasons for this evolution is the realization of a worldwide standard. This dream will not be accomplished by 3G systems, since they will be (more or less) characterized by incompatible standards. These limitations should be overcame by fourth-generation (4G) cellular systems that will create a unique system integrating different access technologies [97]-[99]: a mobile terminal served by a local wireless LAN should be able to roam in a cellular system when leaving a building and moving in a city.

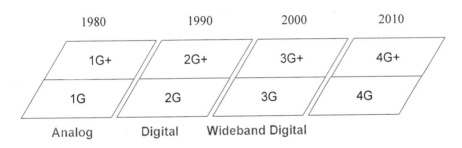

Fig. 64: Wireless systems evolutions towards 4G.

One of the major problems with mobile communication systems is to conjugate user mobility and high-bit-rate support. The picture in Fig. 65 compares the characteristics of different technologies.

Every real 3G system built so far has an absolute maximum speed of 384 kbit/s downstream, 64 kbit/s upstream. A first evolution towards higher capacity is represented by the *High Speed Downlink Packet Access* (HSDPA), a recently standardized improvement for WCDMA, introduced in the Release 5 by 3GPP. HSDPA adopts a multi-modulation technique to provide up to 10 Mbit/s. However, 3G systems will be able to support broadband services only for a limited number of low-mobility users.

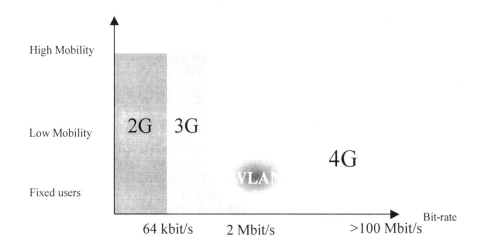

Fig. 65: Mobility vs. capacity trade-off for mobile communications.

Some new aspects that will characterize 4G systems are listed below according to the European standpoint:

- Seamless integration of different access systems (see Fig. 66); this will require the realization of multi-mode terminals able to integrate (for instance) WLAN and cellular accesses.
- Development of innovative air interface schemes allowing for scalable wireless connectivity;
- Widespread provision of multimedia services and applications with high bit-rates and high QoS levels (see Fig. 67);
- Packet-switched traffic over wireless links;
- All-IP network with end-to-end QoS support.
- Reconfigurability of communication links, according to system load, service demand or standard, radio environment characteristics and user profile. Towards this end, software-defined radio concepts will have a major impact.
- Worldwide roaming using a single handheld device.

The foreseen access technologies to be integrated within 4G systems are UMTS, fixed broadband systems, WLAN, HAPS and satellites.

Each sub-system will have different characteristics in terms of coverage range, bandwidth and delay. More details are shown in Fig. 66.

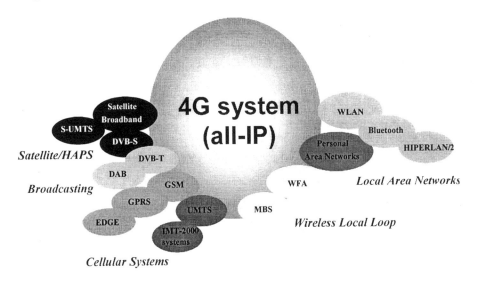

Fig. 66: 4G integrated access systems.

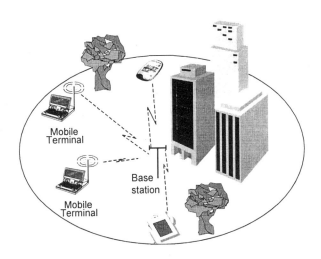

Fig. 67: 4G wireless scenario for high-traffic density urban areas.

4G systems will achieve the *Wireless World*, thus permitting to run a mobile application seamlessly when roaming among different networks.

By means of the IP protocol it will be possible to realize the interoperation of different access systems. Moreover, the adoption of IPv6 will permit to overcome the IP address shortage (thus identifying any device by means of its address). Note that ISPs in Japan have to support IPv6 by 2006.

The 60 GHz band (within the millimeter-wave band) is well suited for very high bit-rates indoor applications with a dense infrastructure, since achievable cell radii are quite limited. A major concern in the use of the 60 GHz band is related to the maximum of the oxygen absorption. Also the 17 GHz band can be considered if larger coverage is needed, provided to accept a lower capacity. Frequency bands below or around 5 GHz should be preferred for wide-area mobile services.

Adaptive modulations and suitable coding will be employed. Moreover, *Orthogonal Frequency Division Multiplexing* (OFDM) scheme is gaining momentum as the transmission technique for providing high bit-rates to mobile users [100]. OFDM is extremely effective in a time-dispersive environment where signals can have many paths to reach their destinations, resulting in variable time delays. With classical modulations, these time delays cause one symbol to interfere with the next one (inter-symbol interference) at high bit-rates. OFDM combats this problem by dividing a radio channel into multiple sub-carriers and by transmitting data in parallel on them. The aggregate throughput is the same, but the data rate of each sub-carrier is reduced (each symbol has a longer duration), thus practically eliminating the effect of inter-symbol interference. The adoption of OFDM requires extremely linear power amplifiers and is facilitated by the efficient implementation of FFT and IFFT algorithms in DSP chips.

Another promising solution to achieve a very high bit-rate wireless access on a short range is given by the *Ultra-WideBand* (UWB) technology. UWB implies the transmission of very short pulses, thus generating a signal with a ultra-wide spectrum. UWB sends low-power (measured in microwatts) coded pulses across an ultra-wide spectrum. UWB does not require an assigned frequency. UWB is an unlicensed communication that would operate in the background of already-occupied frequency bands to provide new applications such as in-building radar and tracking, automobile collision avoidance systems, medical imaging and wireless broadband applications. UWB random, low-power signals can be considered as noise for other transmissions sharing the same bands. UWB has many proponents in the United States, where it has been used for years in defense-related applications. Within the *Institute of Electrical and Electronics Engineers* (IEEE), task group 802.15 is focusing on UWB as *Wireless Personal Area Network* (WPAN). UWB signals can easily pass through non-metal walls, thus the UWB technology is well suited for wireless voice, data and video transmissions inside buildings.

New mobile terminal devices will be needed to support multimedia applications (see Fig. 68).

Fig. 68: Multimedia terminal types.

High bit-rates will be provided to users on the move: downlink transmission peak bit-rate will be around 30 Mbit/s in 2005 and from 50 to 100 Mbit/s in 2010.

As for the possible multimedia applications that will be supported by 4G systems, we can make the following considerations. Successful wireless services are frequently preceded by the growth of a wired demand. Hence, we may expect that the growth in dial-up Internet, and

DSL (*Digital Subscriber Line*) will be precursors for Internet access on the move, cellular data, and 4G broadband wireless. Some Web sites delivers "broadband" video; other Web sites provide some information services (travel information, traffic status, ticket reservation, stock market information, breaking news). These contents will be really accessible through 4G mobile terminals.

Each traffic type will be characterized by suitable high-QoS requirements in terms of delay, loss, jitter. Moreover, we may expect that real-time (isochronous) video traffics and bursty Internet traffics will load 4G networks.

At present, it seems that Japan will be one of the countries where 4G systems will be first implemented, according to the following roadmap:

2005   Complete development of an intermediate system named 3.5G; Establishment of key technologies for 4G systems

2006   Identify the spectrum for 4G at (*World Radio Conference 2005/2006*)

2010   4G services in commercial phase.

5.1   Review on new access technologies

This Section gives a brief survey of the wireless access systems that are foreseen to provide 4G broadband access.

4G-cellular

4G-cellular systems should provide high-speed and should also guarantee high capacity by 2010. The IP protocol will be adopted in the access network.

Multipoint Multichannel Distribution System (MMDS)

An interesting proposal towards the realization of a mobile high bit-rate access is represented by the evolution of fixed wireless systems that already provide multi-megabit speeds. A possible 4G candidate is the MMDS system (in USA) that uses a point-to-multipoint architecture like a cellular network to realize a wireless distribution of *Cable Television* (CATV) traffic in areas where coaxial cables would be impracticable. MMDS adopts a spectrum at about the same frequencies as 3G systems, with a maximum bit-rate lower than 10 Mbit/s. Of course further improvements are needed to increase the bit-rate, to support mobility and to guarantee acceptable transmission also when there is not a clear LOS path. The IEEE is working on an MMDS standard, named 802.16, that will based on OFDM.

The corresponding European solution is represented by the *Local Multipoint Distribution Service* (LMDS), a cellular access technique for high bit-rate data delivery. LMDS operates at millimeter frequencies, typically in 28, 38, or 40 GHz bands, with net data rates up to 38 Mbit/s per user. This technology is able to provide digital video, high-speed Internet data, interactive TV, music and multimedia services.

Wireless LAN (WLAN)

The high bit-rate WLAN technology presently available worldwide is the IEEE 802.11b (Wi-Fi) standard that operates in the 2.4 GHz band and guarantees a maximum information throughput of 7 Mbit/s. Whereas, the 802.11a standard operates in the 5 GHz band, adopts OFDM and reaches a maximum information throughput of 32 Mbit/s.

In Europe, ETSI *Broadband Radio Access Networks* (BRAN) project has developed many standards for high bit-rate radio systems, including: HIPERLAN/2 [101], HIPERLINK and HIPERACCESS. HIPERLAN/2 is almost identical to 802.11a at the physical layer (the same data rates with OFDM at 5 GHz). The difference is that HIPERLAN/2 is designed for both local and wide are networks and includes more advanced QoS and roaming features. HIPERLAN/2 standard has 19 channels spaced by 20 MHz. Each channel will be divided into 52 sub-carriers, with 48 for data and four as pilots that provide synchronization (OFDM scheme).

Mode	Modulation	Code rate	PHY bit-rate	bytes/ OFDM symb.
1	BPSK	½	6 Mbps	3.0
2	BPSK	¾	9 Mbps	4.5
3	QPSK	½	12 Mbps	6.0
4	QPSK	¾	18 Mbps	9.0
5	16QAM	16/9	27 Mbps	13.5
6	16QAM	¾	36 Mbps	18.0
7	64QAM	¾	54 Mbps	27.0

Table 16: HIPERLAN/2 physical layer characteristics.

The problem with WLAN access is that each access point or base station has a very short range, so that WLAN are restricted to a few hotspots (e.g., airport departure lounges, conference centers and hotels). Moreover, the access points have to be connected to the system backbone trough high capacity and very expensive wired lines. ETSI BRAN project is proposing a system that would eliminate the need for these wires. In fact, BRAN has specified a point-to-point version of HIPERLAN/2 called HIPERACCESS, and an entirely new technology called HIPERLINK. HIPERACCESS is used for outdoor fixed links (up to 5 km) at a bit-rate of 25 Mbit/s and operating at frequencies above 11 GHz. HIPERLINK is a standard for high bit-rate links at 155 Mbit/s in the 15 GHz band; it is intended to replace in-building wiring, thus interconnecting HIPERLAN/2 and HIPERACCESS: HIPERLAN/2 is used internally in the building as a WLAN; HIPERLINKs are used to connect HIPERLAN/2 access points to the HIPERACCES distribution system.

In Japan the R&D program called *Multimedia Mobile Access Communication* (MMAC) is developing specifications for four different broadband wireless systems. In particular, we can consider a system, named HisWANa, that is able to transmit up to 30 Mbit/s using a 5.2 band GHz for both outdoor and indoor environments (802.11a, HIPERLAN/2 and HiSWANa standards use nearly identical adaptive modulation and coding techniques at the physical layer; the main differences between 802.11a, HiperLAN/2 and HiSWANa are at the MAC layer). Another system, named *Wireless Home-Link*, will permit transmissions up to 100 Mbit/s using the SHF and other bands (3-60 GHz) in order to interconnect PCs and audiovisual equipment.

HAPS

HAPS systems represent a very interesting solution to provide high bit-rate 4G multimedia services to urban areas. Interconnections with many HAPSs are obtained by optical intercommunication links. A 600 MHz bandwidth in a 48/47 GHz band has been allocated for the fixed services of HAPS. More details on currently planned HAPSs are given in the previous Section 4.3 of Part I.

Satellites

Multimedia satellite systems will provide 4G services to users worldwide. Towards this end, the *Digital Video Broadcasting - Satellite* (DVB-S) standard represents an interesting solution [94]. DVB standards are based on MPEG-2. DVB-S, is the oldest of the DVB standards family. DVB-S is designed to cope with the full range of satellite transponder bandwidths and services. Video, audio and other data are inserted into fixed-length MPEG transport stream packets. *Quadrature Amplitude Modulation* (QAM) is used in DVB-S: the system is based on 64-QAM, but it also allows lower- and higher- level modulations, depending on the trade-off between data capacity and robustness. For instance, an 8 MHz channel with 64-QAM achieves a capacity of 38.5 Mbits/s, the same capacity of a medium-power 36 MHz classical transponder.

5.2   4G view from EU research projects

Many EU R&D projects have addressed the study of 4G systems in RACE II (*Research and Development in Advanced Communications Technologies in Europe*), ACTS (*Advanced Communication Technologies and Services*) and IST (*Information Society Technologies*). In particular, European research on future mobile communications (i.e., 3G and further) started within RACE I (1989) and RACE II (1991). Research on 4G continued in the ACTS program (1994-1999) and in the IST program (1999-2002).

The original vision of 4G systems in RACE II projects was that of new complementary systems able to provide low latency, guaranteed QoS, and a bit-rate of 155 Mbit/s to mobile users. These systems were addressed by the *Mobile Broadband Systems* (MBS) project. MBS thus became the acronym to identify 4G systems. The MBS prototype proved in 1995 the feasibility for a data rate of 34 Mbit/s for both indoor and outdoor mobile environments. MBS physical layer was based on TDMA and higher layers were based on ATM. The MAC protocol adopted by the MBS system was the *Dynamic Slot Assignment++* (DSA++) scheme [102]-[104].

The aims of ACTS Projects (4th framework program) were as follows: implementation aspects, experimentation and validation of equipment and systems in order to properly evaluate the potentiality of future communication services.

In ACTS, six projects focused on 4G systems:

- SAMBA (*System for Advanced Mobile Broadband Applications*),

- MEDIAN (*Wireless Broadband CPN/LAN for Professional and Residential Multimedia*),

- WAND (*Wireless ATM Network Demonstrator*),

- AWACS (*ATM Wireless Access Communication System*),

- ACCORD (*ACTS Broadband Communication Joint Trials and Demonstrations*),

- SECOMS (*Satellite EHF Communications for Mobile Multimedia*),

- SORT (*Software Radio Technology*),

- CABSINET (*Cellular access to broadband services and interactive TV*).

These projects addressed both terrestrial and satellite mobile communication systems. The followed approach was essentially that of promoting the realization of *Wireless ATM* (WATM) systems that, at that times, was seen as the real achievement of 4G systems. In particular, we can consider the following short descriptions of these projects.

SAMBA developed a trial for high bit-rate full-duplex transmissions at 34 Mbit/s in the 40 GHz band.

MEDIAN project focused on the definition of a high-speed WLAN supporting data rates up to 150 Mbit/s in indoor environments using the 60 GHz band.

The WAND project demonstrated a high-speed, wireless access systems for ATM networks. This project focused on the following aspects:

- Specification of a wireless access system for ATM networks that maintains the service characteristics and benefits of ATM networks;
- Realization of a WATM demonstrator in the 5 GHz band, allowing a user transmission capacity up to 20 Mbit/s;
- Promotion of ETSI standardization towards WATM.

The WAND project contributed to the definition of the HIPERLAN/2 standard. The MAC layer of WAND was based on an innovative protocol named *Mobile Access Scheme based on Contention and Reservation for ATM* (MASCARA) with *Prioritized Regulated Allocation Delay Oriented Scheduling* (PRADOS) [105].

AWACS project focused on the realization of a wireless local loop. It built a demonstrator operating in the 19 GHz band with a transmission rate of 34 Mbit/s in a TDD mode.

ACCORD project aimed to harmonize the major findings of four ACTS projects focused on different broadband communication systems in order to integrate them (i.e., multiple satellite and terrestrial system components with complementary characteristics).

SECOMS aimed at defining the system elements and related technologies for future advanced satellite services at Ka and EHF bands. Geostationary satellites were considered. Services were considered at a rate of nx64 kbit/s, compatible with the ISDN primary rate access. The 20/30 GHz and 40/50 GHz bands were adopted for the first- and second-generation systems respectively.

SORT project dealt with radio access reconfigurability, a promising technological solution to achieve flexible, multi-band and multi-mode mobile terminals. SORT focused on the demonstration of efficient software programmable radios, mainly referring to 3G systems.

CABSINET project aimed to deliver new services such as entertainment, healthcare and education to users either currently unable to receive fiber-optic capability in the home, or for whom such provision is uneconomical. The LMDS system foreseen by CABSINET was based on local repeaters close to user premises at 5.8 GHz (microcells); repeaters were linked to a central station at 40 GHz (macrocells). FDMA/TDMA was used for macrocells, whereas TDMA over DS-CDMA was adopted for microcells.

After ACTS projects, further improvements have been introduced towards the definition of 4G systems by IST Projects (5th framework program) [106]. In particular, OFDM has been considered as the most suitable physical layer technique and ATM has been replaced by IP at the network layer. Among IST projects, we may consider those described below.

- BRAIN (*Broadband Radio Access for IP Networks*) project is working on an IP-based wireless access network that implements micro-mobility and supports QoS. BRAIN permits that different air interfaces (i.e., GSM, UMTS, HIPERLAN/2) be integrated in the same system though the IP protocol. Mobility functions are envisaged in the BRAIN approach to support handoffs within a network and between different networks.

- WINE GLASS (*Wireless IP NEtwork as a Generic platform for Location Aware Service Support*) project uses state-of-the-art UTRAN to interconnect directly to an enhanced Mobile IP v6 backbone through a standard Iu interface. This approach allows a significant simplification of procedures for session management, mobility and authentication.

- MIND (*Mobile IP-based Network Developments*) project focuses on both the mobile IP issue and the end-to-end QoS support in the light of 4G networks.

Chapter 1: General Concepts on Radio Resource Management

The efficient management of radio resources is a crucial aspect for the air interfaces of all the wireless systems described in the first Part of this book. In fact, radio resources are scarcely available, costly and error prone. Hence, suitable techniques must be adopted to guarantee high capacity of simultaneous users and the fulfillment of QoS levels for the different traffic classes. This is the typical task of layer 2 protocols of the OSI Reference Model, including *Medium Access Control* (MAC) and *Usage Parameter Control* (UPC) protocols. All these protocols are here jointly identified with the name *Radio Resource Management* (RRM). We refer to a centralized RRM scheme, where a base station (or an RNC in the UMTS system or an *Access Point* in a wireless system or a satellite with on-board processing capabilities) decides the allocation of *resource quanta* to the different traffic flows. In addition to this, we consider only the case where packet-switched traffic has to be managed, since it can provide superior performance in multiplexing bursty traffic sources.

We distinguish between RRM management tasks for uplink and those for downlink resources. These two tasks are distinct in *Frequency Division Duplexing* (FDD) air interfaces, whereas they can be jointly performed in *Time Division Duplexing* (TDD) cases. Of course, TDD air interfaces have the advantage that the resource manager can dynamically balance uplink and downlink resources depending on their respective traffic loads. This is particularly important in the presence of Web traffics loading the air interface, since upstream/downstream loads are typically in the ratio of 1:10 [107].

Uplink traffics are from the *Mobile Terminals* (MTs) to the *Base Station* (BS); whereas downlink traffics are from the BS to MTs. Since we focus on OSI layer 2, we refer here to uplink and downlink traffics of sessions already opened by higher OSI layers. Sessions are admitted depending on their traffic loads and QoS requirements and verifying whether they can be supported in the presence of already active sessions. This is the typical task of *Call Admission Control* (CAC) protocols.

We can consider that MTs have to transmit/receive traffics belonging to different traffic classes (i.e., conversational, streaming, interactive and background, according to the common classification made in 3G systems).

The resource space managed by the RRM scheme (MAC plus UPC protocols) depends on the multiple access technology adopted on the air interface, that is frequency division, time division, code division or hybrid approaches.

The BS performs the management of uplink traffics on the basis of transmission requests coming from MTs. A signaling method must be used for the exchange of this information:

-   Transmission requests of inactive MTs (i.e., MTs that have opened a session, but that currently have no traffic to be sent to the BS) are transmitted by means of a *random access scheme* carried out on a given portion of resources of the air interface.

-   Transmission requests of active MTs may be sent to the BS as *piggybacked messages* in their uplink traffic flows.

All the requests coming from MTs are put in suitable service queues at the BS in order to reproduce virtually the behavior of the transmission queues of each MT.

As for downlink, the BS has transmission queues that directly collect the traffics to be managed for the MTs in its cell.

Hence, we may note a first difference in the management of uplink and downlink resources: the BS immediately and exactly knows the transmission needs for downlink traffics; whereas, the BS knows with variable delays (due to both the contention phase in the random access scheme and the availability of uplink transmissions for the piggybacking technique) the transmission needs for the traffic flows coming from MTs.

According to the per-flow QoS support of the DiffServ architecture proposed for the Internet [46], an efficient solution to manage multimedia traffics (each with specific QoS requirements) is to use a different queue for each traffic class. The queue service order and the amount of service provided to each queue are tasks of MAC and UPC protocols.

The MAC protocol is implemented in a *scheduler* at the BS that decides the amount of traffics to be served for the different queues. The UPC protocol typically is implemented in a *policer* at the BS to share fairly the resources for the transmissions of the MTs. Hence, MAC and UPC jointly operate for the complex task of allocating resources to the different traffics on the air interface.

We can consider the following taxonomy for MAC protocols [108]:

1. *Fixed access* protocols that grant permission to send only to one MT at once, avoiding collisions of messages on the shared medium. Access rights are statically defined for the MTs.

2. *Demand-adaptive* protocols that grant the access to the network on the basis of requests made by the MTs. This class encompasses reservation and token-based schemes.

3. *Contention-based* protocols that give transmission rights to several MTs at the same time. This policy may cause two or more MTs to send simultaneously so that their messages collide on the shared medium.

Typical examples of fixed access schemes are FDMA, TDMA and CDMA techniques. Many contention-based schemes have been proposed that represent evolutions of both the classical *Slotted Aloha* technique and the *Packet Reservation Multiple Access* (PRMA) protocol [109]. Finally, typical examples of demand-adaptive schemes are the *Reservation Resource Allocation* (RRA) protocols in [110].

Fixed access schemes are not efficient with bursty traffics, because they cannot adapt to varying traffic conditions. Moreover, contention-based schemes are not adequate to manage correlated and heavy traffics. An advantage of demand-adaptive protocols over most contention-based ones is the existence of an upper bound on the transmission delay. The problem with MAC schemes it that they were initially conceived to support only one traffic class, whereas it is important now to develop

RRM protocols able to manage multimedia traffics with different characteristics in terms of burstiness, QoS, load, etc. Interesting solutions are represented by both demand-adaptive and contention-based solutions.

RRM schemes must be defined according to the following requirements:

- Scheduling among traffic classes taking into account different priorities and scheduling within a traffic class on the basis of urgency, QoS and other parameters.

- Provision of a fair service among the traffic sources within a class.

- Management of several traffic classes guaranteeing their different service priorities and their specific QoS levels.

- Optimal MAC techniques must be adaptive to traffic load conditions, higher-layer protocol behaviors (including applications) and radio channel status in order to guarantee the maximum throughput.

- To achieve a high utilization of radio resources.

- To allow a prompt access to resources, an essential prerequisite for supporting real-time and interactive traffics.

- To guarantee the protocol stability, that is a correct protocol behavior. In particular, we consider the ability to manage correctly the access requests of the different MTs.

Typical scheduling techniques for the packets in a queue are [111]:

- *First Input First Output* (FIFO): packets are managed according to their arrival instants.

- *Earliest Deadline First* (EDF): packets are served according to an urgency criterion: each packet has a deadline to be transmitted; the packet with the shortest residual life is transmitted. Such scheme requires the dynamic management of the buffer for each traffic class when we have to serve packets with different deadline values (otherwise EDF practically becomes a FIFO scheme).

- *Round Robin* (RR): resources are cyclically assigned to the different requests[8]. There are also different versions of RR. For example, we can consider *Weighted Round Robin* (WRR), where resources are cyclically assigned with an amount depending on the weight assigned to each traffic. WRR is a practical implementation of the *Processor Sharing* (PS) technique, an unrealizable (ideal) scheduler that assumes a bit-by-bit round robin scheme among all the messages to be transmitted for achieving the maximum fairness level in sharing resources.

The EDF scheme is quite appropriate for the management of real-time traffics. RR schemes are valid solutions for transmitting interactive traffics, since RR techniques can guarantee a maximum delay in the service of a source (correspondingly, a minimum bit-rate), thus allowing a sufficiently fast exchange of data between MTs and BS. Moreover, RR schemes are well suited for managing the transmissions of messages having a high ratio between variance and mean message length. In fact, it has been proved in [5] that the ideal PS scheme is more convenient[9] than the FIFO discipline, if the transmission time distribution of messages, τ, fulfills the following condition:

$$\frac{E[\tau^2]}{2E[\tau]^2} > 1 . \tag{1}$$

Such condition is typically verified by heavy-tailed distributions, the typical case for the length of objects downloaded from the Internet.

The RR scheme guarantees that a request/message be served within a maximum delay; this is not possible with the FIFO scheme. Hence, the RR approach permits to reduce the high-value percentile of the transmission delays with respect to the FIFO solution.

More refined scheduling schemes must base their transmission priorities not only on the deadline of packets (if any), as in the EDF

---

[8] Note that *polling schemes* are particular cases of round robin techniques if applied to a distributed queue in spatially dispersed MTs. Hence, polling can be used for uplink transmissions, but entails some inefficiencies due to the delays for switching the service for an MT to another (protocol overhead).

[9] In terms of the mean transmission delay conditioned on messages of a given length.

case, but also on the buffer congestion (a more congested buffer for a source could imply that more resources need to be destined to transmit its packets), channel quality (including bit error rate and *Carrier-to-Interference* ratio experienced by each MT). In addition to this, different priority levels must be guaranteed for traffics belonging to distinct classes. Accordingly, an interesting requisite for RRM schemes (and related scheduler) is that real-time traffic delays are not affected by the presence of low-priority data traffic flows (QoS insulation among different traffic classes). Hence, an appropriate prioritization scheme is needed.

Typical UPC schemes are [21]:

- ***Leaky bucket shaper***: it is used at the traffic source to control the traffic transmission at a constant rate defined at the session set-up. The output traffic from this regulator as a lower burstiness value than the input traffic.

- ***Token bucket policer***: the token is a permission to transmit. The BS uses a bucket for each source, where tokens are put at rate of r tokens/s (corresponding to the mean bit-rate defined at the session set-up with the MT). The bucket has a capacity of B tokens (see Fig. 1) that correspond to the largest burst an MT can transmit. If the bucket is full, newly arriving tokens are discarded. When an MT is enabled to transmit, the number of tokens in its bucket regulates the maximum number of packets that can be sent.

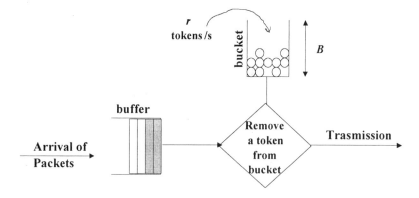

Fig. 1: Token bucket regulator applied to the management of a transmission queue.

- ***Fuzzy logic-based policer*** [112]: this is a complex traffic regulator operated by the BS for each traffic source on the basis of policing rules defined according to the experience and based on the input coming from the traffic source status itself and the contractual parameters including the required QoS levels.

A simple example of joint use of scheduler and policer is as follows: adoption of the RR scheduler to serve the different sources with traffics in their queues and a token bucket policer to determine the number of packets transmitted from each queue per cycle (a more detailed description of this approach will be given in the following Chapter 6).

In order to summarize all the above concepts related to RRM performed by the OSI layer 2 protocols at the BS, we may refer to Fig. 2 that describes the management of different traffic classes by different queues by means of a scheduler and a policer.

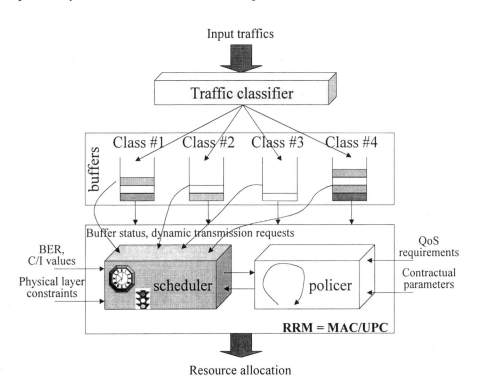

Fig. 2: Architecture of a general scheme to manage traffics at layer 2 of the BS.

As shown in Fig. 2, the scheduler and the policer jointly cooperate in defining resource allocations on the basis of transmission requests, contractual parameters, QoS levels and estimated channel status conditions (BER values, interference levels). Moreover, RRM for 3G cellular systems have also to define physical layer parameters such as transmission power and processing gain values. An important physical layer constraint is related to the maximum transmission power level. In particular, we have to consider the maximum MT transmission power in uplink and the maximum BS transmission power in downlink.

Referring to the future scenario of 4G systems, where different mobile networks will cooperate to provide a seamless coverage and (possibly) a seamless QoS provision, new RRM schemes need to be defined to account for the layered (hierarchical) cellular coverage. In such scenario, two approaches are available: *horizontal integration* and *vertical integration*. In the first case, a centralized management of resources belonging to the different layers is required for an effective support of mobile users. In fact, it is important to define criteria according to which an MT uses the resources of one layer (e.g., micro-cells) or of another (e.g., macro-cells) [113]. Whereas, the vertical approach refers to the realization of an adaptive radio resource management that takes scheduling decisions on the basis of traffic load conditions, radio propagation conditions and higher-layers behaviors; in this case, OSI layer 2 choices on the air interface also depend on the protocols of the other layers.

The following Chapters describe some proposed RRM strategies for different air interfaces: GPRS, WCDMA, UTRA-TDD, WATM based on TDMA, satellite air interface based on TDMA. Performance evaluations are also carried out by means of simulations on the basis of the traffic models described in the next Chapter. In addition to this, some theoretical methods will be described that can be used to analyze the performance of different RRM protocols in the presence of multimedia packet data traffics. Finally, protocols allowing the Internet access through mobile devices will be investigated with a special attention to the modifications to be introduced to account for the air interface characteristics.

Chapter 2: Traffic models

This Chapter describes the models for packet data traffics that can be adopted to evaluate the performance of RRM schemes. Moreover, we show also simple radio channel models that can be adopted to simulate the traffic management in mobile environments.

We start below by surveying traffic models for voice sources, video sources and Web browsing sources. For the sake of simplicity, we will consider sources as producing traffic directly offered to the radio resource manager at the OSI layer 2 (even if some *overhead* should be included to account for intermediate protocol layers). Hence, this traffic has to be packetized according to the layer 2 packet format (i.e., payload). Let T_{pkt} denote the transmission time of this packet.

2.1   Voice sources

We assume that each voice source (conversational traffic class) uses a *Speech Activity Detector* (SAD) to distinguish between talking and silent phases [109]. During a talkspurt (ON state), a voice source produces a constant bit-rate R that depends on the adopted codec. No traffic is generated in a silent pause (OFF state). Sojourn times in ON and OFF states are exponentially distributed with mean values $1/\omega = 1$ s and $1/\zeta = 1.35$ s (see Fig. 3), respectively. Hence, the voice activity factor ψ_s is equal to *Prob.*{ON state} = $(1/\omega)/(1/\omega+1/\zeta) = 0.425$.

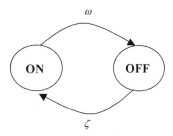

Fig. 3: Voice source ON-OFF traffic model.

The *burstiness degree*, B, of the traffic produced by a source is equal to the ratio of the peak bit-rate and the mean bit-rate. In the case of the voice ON-OFF source, B $= R/(\psi_s R) = 1/\psi_s$. This burstiness value represents the maximum (ideal) multiplexing gain that can be achieved when transmitting different ON-OFF voice sources (of this type) on a link with capacity R.

During a talkspurt the bits produced by a voice source are packetized. Voice traffic is real-time. Hence, voice packets have stringent delay requirements: they are discarded if they experience an access delay greater than D_{smax}, (assumed equal to 32 ms [109]). A typical QoS requirement is that the *speech Packet Dropping probability*, $P_{drop,s}$, must be lower than 1% to preserve voice quality.

2.2   Video sources

We consider two traffic models for constant-quality variable bit-rate video sources to be used for both conversational (e.g., videoconference) and streaming traffic classes (e.g., video on demand). In the first model, the real-time traffic produced by a video source can be considered as the aggregated output of M independent *minisources*, each alternating between OFF and ON states [105],[114]. Each minisource in the ON state produces traffic at the constant rate of A bit/s; whereas in the OFF state no traffic is generated. Time is supposed to evolve in discrete steps, named slots, T_s, that can (for instance) correspond to the time to transmit a packet produced by this source (i.e., T_{pkt}). The time intervals (in slots) spent in ON and OFF states are geometrically distributed with parameters $\alpha = 1/p$ and $\beta = 1/q$, where p (and q) is the mean time spent in ON (and OFF) in slots. A minisource in the ON (or OFF) state makes a transition towards the OFF (or ON) state at the end of a slot with probability α (or probability β). The minisource activity factor is $\psi_v = p/(p + q)$. We have considered that in a slot at most one minisource can make a transition from ON to OFF or vice versa. Hence, there are not sudden traffic variations for a video source (i.e., videoconference case). Hence, the whole video traffic source can be modeled through the discrete-time Markovian modulating process (D-MAP) described in Fig. 4. Parameters p, q and A of a minisource can be obtained as follows [105],[114]:

$$A = \frac{\mu}{M} + \frac{\sigma^2}{\mu} \left[\frac{\text{bit}}{\text{s}} \right], \quad q = \frac{1}{aT_s} \left(1 + \frac{M\sigma^2}{\mu^2} \right) \, [\text{s}],$$

$$p = \frac{1}{aT_s} \left(1 + \frac{\mu^2}{M\sigma^2} \right) \, [\text{s}] \tag{2}$$

where μ is the mean bit-rate produced by a source, σ^2 is the variance of the bit-rate produced by a source and parameter a characterizes the slope of the auto-covariance function[10] of the bit-rate produced by a source (if a decreases, a more correlated traffic generation process is obtained).

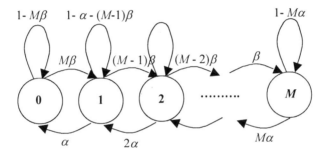

Fig. 4: Modulating process for the packet generation of a VT (D-MAP model): transitions occurs on a slot basis.

It is easy to show that the state probability of the modulating process in Fig. 4 is binomial:

$$Prob.\{state = i\} = \binom{M}{i} \psi_v^i \left(1 - \psi_v \right)^{M-i} \quad i = 0, ..., M . \tag{3}$$

In this model, a bit-rate iA is produced when the video source is in the state i. The mean bit-rate produced is $\mu = M\psi_v A$ bit/s and the maximum bit-rate is MA. Therefore, the burstiness degree for a video source is B $= 1/\psi_v$.

---

[10] The auto-covariance function $C(\tau)$ for the bit-rate generated by a video source depends on time τ as follow: $C(\tau) = \sigma^2 e^{-a\tau}$. This exponentially decaying auto-correlation function is characteristic of a short-range dependent traffic.

We can use $5 \leq M \leq 10$ to model a single video source. A typical choice for the parameter values is as follows [105],[114]: $\mu = 512$ kbit/s, $\sigma = 256$ kbit/s, $M = 10$ minisources/video and $a = 3.9$ s^{-1} (also lower a values can be used for videoconference). With these values, the minisource activity factor ψ_v is equal to 0.28 and the burstiness B is 3.571. Finally, the bits generated by a video source are packetized according to the packet payload. We consider that each video packet produced must be transmitted within a deadline D_{vmax} with typical values in the range 50-150 ms.

A video traffic source based on the modulating process as in Fig. 4 can be implemented as follows: as soon as the source enters the state i with leaving probabilities $i\alpha$ and $(M - i)\beta$, we can generate two samples t_{incr}, t_{decr} from geometrically distributed random variables with mean values $1/(i\alpha)$ and $1/[(M - i)\beta]$, respectively. Hence, the sojourn time for state i in T_s units can be determined as $t_{sog} = \min\{t_{incr}, t_{decr}\}$; correspondingly, the transition accomplished at the end of interval t_{sog} is towards state $i - 1$, if $t_{decr} \equiv \min\{t_{incr}, t_{decr}\}$; otherwise, the transition is towards state $i + 1$. During t_{sog} we have a source generating traffic at a constant bit-rate equal to iA.

We present below another video traffic model, where the traffic produced by a source is the superimposition of M (typically 5) independent minisources, each alternating between OFF and ON states [115]. Each ON minisource generates a packet every 6 ms corresponding to a bit-rate $A = 64$ kbit/s. ON and OFF sojourn times are exponentially distributed with mean values $1/\beta = 33$ ms and $1/\alpha = 67$ ms, respectively. The video minisource activity factor ψ_v results to be 0.33 and the burstiness degree B is about equal to 3. Every 40 ms, the generated packets are collected to form a *video frame* where all the packets have the same deadline (= 40 ms).

For both the video traffic models, we adopt the QoS requirement that the *video Packet Dropping Probability, $P_{drop,v}$* must be lower than 10^{-4} [116].

2.3   Web browsing sources

We use the following model for a traffic source producing Internet browsing traffic (for uplink and downlink, with suitable scale factors).

Referring to a given Web session, a Web source (interactive traffic class) alternates between a *packet call state* during which datagrams are generated and a *reading time state* where no traffic is produced. This is a simplified model derived from [53],[117]. The number of datagrams per packet call is geometrically distributed with expected value $m_{Nd} = 25$. The datagram interarrival time and the reading time are exponentially distributed with mean values $m_{Dd} = 0.5/q$ s (where $q = 1$, 2, 3, ... is used to modulate the burstiness degree of the traffic produced by this source) and $m_{Dpc} = 12$ s, respectively. The parameters of this model can be 'scaled' to cover also different Internet surfing scenarios, such as *Wireless Application Protocol* (WAP) browsing through 2G or 2.5G mobile phones [118]. Moreover, several concurrent Internet sessions for a given user could be generated by considering the superposition of many traffic sources of the type described here.

The continuous-time modulating process associated with a Web traffic source is depicted in Fig. 5 (we will prove later that this is a Markov chain).

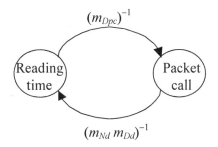

Fig. 5: Web source modulating process.

The activity factor for this source is $\psi_w = m_{Nd} m_{Dd} / (m_{Nd} m_{Dd} + m_{Dpc})$. The mean arrival rate of datagrams is $\lambda_w = \psi_w / m_{Dd}$.

Each datagram has a random length in bytes $l_{wbyte} = \lfloor x \rfloor$, where $\lfloor . \rfloor$ denotes the *floor function* and x is a random variable with the following truncated Pareto *probability density function* (pdf) with parameters $\upsilon = 1.1$, $k = 81.5$ bytes and $m = 66666$ bytes (the maximum IPv4 datagram length is about 65 kbytes, even if some LANs require much shorter datagrams) [53]:

$$pdf(x) = \frac{\upsilon k^{\upsilon}}{x^{\upsilon+1}}\left[u(x-k) - u(x-m)\right] + \left(\frac{k}{m}\right)^{\upsilon}\delta(x-m) \qquad (4)$$

where $u(.)$ is the *unitary step function* and $\delta(.)$ is the *Dirac delta function*.

Without truncation, the Pareto pdf may have infinite mean and/or infinite variance, depending on the υ value.

The mean value of x can be obtained as follows:

$$E[x] = \frac{\upsilon k - m\left(\frac{k}{m}\right)^{\upsilon}}{\upsilon - 1} \quad . \qquad (5)$$

With the above values the mean datagram length is 481 bytes.

Table 1 shows the burstiness degree B $= 1/\psi_w$ and the mean bit-rate $\lambda_w E[x]$ as functions of parameter q.

Parameter q value	Burstiness degree, B	Mean bit-rate, $\lambda_w E[x]$
1	1.31	5.83 kbit/s
2	1.64	9.38 kbit/s
3	1.96	11.78 kbit/s
4	2.27	13.50 kbit/s
5	2.79	14.80 kbit/s
6	2.92	15.81 kbit/s
7	3.23	16.62 kbit/s
8	3.57	18.00 kbit/s

Table 1: Characteristics of a Web traffic source for different q values.

The Pareto distribution considered in (4) is heavy-tailed, meaning that it has a high mean square value as compared to the square of the mean value. This is pictorially evident in Fig. 6, where the Pareto pdf (4) is compared with an exponential distribution with the same mean value.

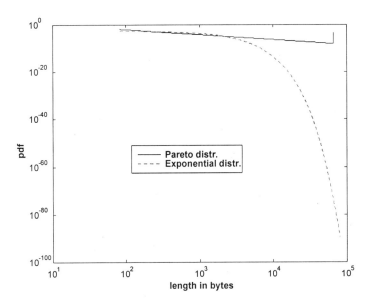

Fig. 6: Comparison between the Pareto pdf of datagrams and the corresponding exponential distribution.

Datagrams are packetized according to the layer 2 format. From (4) we obtain the following probability mass function of the datagram length, l_w, in packets with payload L_p bytes (assuming that $L_{w,min}$ packets are needed to transmit the shortest datagram of $k = 81.5$ bytes):

$$Prob.\{l_w = j \text{ packets}\} = \begin{cases} 1 - \left(\dfrac{k}{jL_p + 1}\right)^\upsilon, & j = L_{w,min} \\[2ex] \left(\dfrac{k}{(j-1)L_p + 1}\right)^\upsilon - \left(\dfrac{k}{jL_p + 1}\right)^\upsilon, & L_{w,min} < j < L_{w,max} \\[2ex] \left(\dfrac{k}{(j-1)L_p + 1}\right)^\upsilon, & j = L_{w,max} \end{cases} \qquad (6)$$

where $L_{w,min} = \lceil k/L_p \rceil$, $L_{w,max} = \lceil m/L_p \rceil$, and symbol $\lceil . \rceil$ denotes the *ceiling function*.

According to the above probability mass function, we can derive mean and mean square datagram lengths in packets, parameters of significant importance that characterize the queuing performance of a transmission buffer. Let $E[l_w]$ denote the mean datagram length in packets. The layer 2 traffic load ρ_w in Erlang produced by a Web source of this type can be expressed as:

$$\rho_w = \lambda_w E[l_w] T_{pkt} \quad [\text{Erlang}] \ . \tag{7}$$

Typical QoS parameters for Web traffic sources are both the mean and the 95-th percentile of the packet or of the datagram transmission delay.

Of course the traffic produced by a Web source depends on the TCP behavior that may slow down transmissions due to the *congestion avoidance algorithm* in the presence of high delays or packet errors on the air interface. In such cases, the ON-OFF Web source traffic behavior described here is altered due to slowed interactions. This *backlog effect* is not considered in the Web traffic model proposed here [53].

It is easy to show that each Web traffic source proposed here produces a *two-state Markov Modulated Poisson arrival Process* (2-MMPP) of datagrams, thus allowing the adoption of a wide literature for analyzing queue behaviors. Indeed, the sojourn time in the reading time state is exponentially distributed and no traffic is generated. Moreover, the sojourn time in the packet call state results to be the sum of a geometrical number of exponentially distributed interarrival times. The *Probability-Generating Function* (PGF) of the number of datagrams per packet call can be expressed as [5]:

$$N(z) = \frac{z}{m_{N_d}\left[1 - \left(1 - \dfrac{1}{m_{N_d}}\right)z\right]} \ . \tag{8}$$

The *Laplace-Stieltjes transform* (LST) of the datagram interarrival time in the packet call state is:

$$M(s) = \cfrac{1}{m_{D_d}\left(\cfrac{1}{m_{D_d}} + s\right)} \quad . \tag{9}$$

Finally, the time spent in the packet call state is the sum of a random number of datagram interarrival times. Thus, the corresponding LST is obtained from the composition of $N(z)$ and $M(s)$ as follows:

$$N[M(s)] = \cfrac{1}{m_{N_d} m_{D_d}\left(\cfrac{1}{m_{N_d} m_{D_d}} + s\right)} \quad . \tag{10}$$

Comparing (10) with (9), we note that the sojourn time in the packet call state is exponentially distributed with mean value $m_{Lpc} = m_{Nd} m_{Dd}$. Finally, recalling that the datagram interarrival time in the packet call state is exponentially distributed, we can conclude the proof of the 2-MMPP arrival process produced by each Web browsing traffic source.

2.4   Self-Similar traffic sources

We consider a special traffic situation that is the aggregation of many Web browsing sources. This could be the case where the IP traffic produced by many MTs on a local area network is conveyed via a wireless link to an access point. A complex aggregated traffic model has to be considered, as described below.

A *self-similar traffic* trace shows structural likeness for a wide range of time aggregations. Moreover, a *Long-Range Dependent* (LRD) arrival process has non-summable auto-correlation function, thus causing high delays in the networks. Long-range dependence implies that traffic smoothly evolves with time, thus having very long traffic peaks. However, the most surprising characteristic of IP traffic is self-similarity. A self-similar traffic trace shows structural likeness for a wide range of time aggregations. This suggests that fractals are the most appropriate mathematical tool to describe certain aspects of IP-based networks.   This was the astonishing discovery of Bellcore researchers in the late 1980s and early 1990s. Hence, traffic burstiness in IP networks cannot be averaged out, thus posing significant problems in dimensioning link capacities to fulfill given QoS requirements [119].

A stochastic process $Z(t)$, $t \in R^+$, is *strictly self-similar* (ss) with Hurst parameter $H \in [0.5, 1)$, if for a certain $\alpha > 0$ it satisfies [120]:

$$Z(\alpha t) \overset{d}{=} \alpha^H Z(t) \ . \tag{11}$$

where symbol "$\overset{d}{=}$" means equality of all the finite-dimensional joint distributions.

We denote with $X_n = \{X_0, X_1, \ldots\}$ a semi-infinite segment of a second-order stationary stochastic process[11]. Let $\mu = E(X_n)$ and $x_n = X_n - \mu$. The X_n process has normalized auto-correlation function as:

$$r(k) = \frac{E[X_n X_{n+k}]}{E[X_n^2]} \ . \tag{12}$$

Process X_n is *exactly second-order self-similar* (es-ss) with Hurst parameter $H = 1 - \beta/2$, $0 < \beta \leq 1$, if its auto-correlation function satisfies:

$$r(k) = \frac{1}{2}\left[|k+1|^{2-\beta} - 2|k|^{2-\beta} + |k-1|^{2-\beta}\right] \ . \tag{13}$$

We introduce the aggregated process $X_n^{(m)}$:

$$X_n^{(m)} = \frac{1}{m}\left[X_{nm} + \ldots + X_{nm+m-1}\right] \tag{14}$$

where $n \in I_0 = \{0, 1, \ldots\}$ and $m \in$ in $I_0 - \{0\}$.

Let $r^{(m)}(k)$ denote the normalized auto-correlation function of $X_n^{(m)}$. The X_n process is *asymptotically second-order self-similar* (as-ss) with

---

[11] X_n has mean value and normalized auto-correlation function that do not depend on $n \in I_0 = \{0, 1, \ldots\}$.

Hurst parameter $H = 1 - \beta/2$, $0 < \beta \le 1$, if the auto-correlation function of $X_n^{(m)}$ satisfies the following equality:

$$\lim_{m \to +\infty} r^{(m)}(k) \equiv r(k) \ .$$
(15)

A second-order stationary process X_n is said to be LRD if its normalized auto-correlation function is such as:

$$r(k) \propto c_r |k|^{\alpha-1}, \quad k \to +\infty, \quad \alpha \in (0,1), \quad \alpha = -1+2H \ .$$
(16)

A typical self-similar and LRD traffic source is given by the M/Pareto model [121]. M/Pareto traffic is generated as Poisson arrivals of overlapping bursts. Let λ denote the mean arrival rate of bursts. Hence, the number of messages arrived in an interval of length t, $A(t)$, is Poisson distributed as:

$$Prob.\{A(t) = n\} = \frac{(\lambda t)^n}{n!} e^{-\lambda t} \ .$$
(17)

The packet arrival process is constant for the duration of the burst with rate r packets/s. The duration of each burst is a random variable according to a Pareto distribution with complementary distribution as detailed below:

$$Prob.\{X > x\} = \begin{cases} \left(\dfrac{x}{\delta}\right)^{-\gamma}, & x \ge \delta \\ 1, & \text{otherwise} \end{cases}$$
(18)

where $1 < \gamma < 2$ (for having the Hurst parameter in [0.5, 1) for a self-similar traffic, as detailed below) and $\delta > 0$.

The mean of X is $\delta\gamma/(\gamma - 1)$ s and its variance is infinite. The mean number of packets within one burst is $r\delta\gamma/(\gamma - 1)$.

This traffic model corresponds to an $M/G/\infty$ system [5] (Poisson arrivals of bursts/General burst duration/infinite bursts can be simultaneously present); hence, a Poisson distribution of the number N

of simultaneously present bursts can be considered with mean value $\rho = \delta\gamma\lambda/(\gamma - 1)$:

$$Prob.\{N = k\} = \frac{\rho^k}{k!} e^{-\rho} . \quad (19)$$

The mean traffic produced by an M/Pareto source is $r\rho$ in packets/s.

Although the Pareto burst length has infinite variance, the variance of the M/Pareto process is finite and can be express as follow [121]:

$$\sigma^2(t) = \begin{cases} 2r^2\lambda t^2\left(\frac{\delta}{2}\left(1 - \frac{1}{1-\gamma}\right) - \frac{t}{6}\right) & 0 \le t \le \delta \\ \\ 2r^2\lambda t^2\left\{ \begin{aligned} &\delta^3\left(\frac{1}{3} - \frac{1}{2-2\gamma} + \frac{1}{(1-\gamma)(2-\gamma)(3-\gamma)}\right) + \\ &+ \delta^2\left(\frac{1}{2} - \frac{1}{1-\gamma} + \frac{1}{(1-\gamma)(2-\gamma)}\right)(t-\delta) + \\ &- \frac{t^{3-\gamma}}{\delta^{-\gamma}(1-\gamma)(2-\gamma)(3-\gamma)} \end{aligned} \right\} & t > \delta \end{cases} \quad (20)$$

In (20) we note that the dominant term is $2r^2\lambda\dfrac{\delta^\gamma t^{3-\gamma}}{(1-\gamma)(2-\gamma)(3-\gamma)}$ for large t; this function is proportional to $t^{3-\gamma}$. Hence, the M/Pareto model generates *asymptotically self-similar* traffic with Hurst parameter $H = \dfrac{3-\gamma}{2}$. The greatest H, the higher the traffic correlation degree and the worse the queuing performance in the presence of an M/Pareto traffic source.

2.5   Data traffic sources

For background sources [17] we assume a classical memoryless traffic model, where message arrivals are Poisson distributed with mean rate λ_d msg/s and message length l_d in packets geometrically distributed with mean value $L_d = E[l_d]$:

$$Prob.\{l_d = n \text{ packets}\} = \frac{1}{L_d}\left(1 - \frac{1}{L_d}\right)^{n-1} \quad n = 1, 2, \dots \quad . \quad (21)$$

The burstiness degree for this source is $B = 1$.

No specific delay requirement has been used for background traffics.

Another model for background email traffic sources proposed in [117] still assumes a Poisson arrival of messages, but with a Pareto length distribution.

Note that bursty packet traffics ($B > 1$) yield higher queuing delays than non-bursty packet traffics ($B = 1$), at a parity of traffic intensity.

2.6   Channel models

This Section deals with radio channel modelization for typical layer 2 simulations.

Radio propagation strongly depends on both the transmission frequency and the environment (e.g., urban area, suburban area, rural area and hilly terrain). For example the free space path loss attenuation L is:

$$L = \left(\frac{4\pi D f}{c}\right)^2 \quad . \quad (22)$$

where D = distance, f = transmission frequency, c = light-speed in the vacuum.

In a mobile environment other laws must be used [122]. A typical path loss model adopted for 3G systems is that of the *outdoor-to-indoor and pedestrian test environment* in [53]:

$$L = 40\log_{10}(D) + 30\log_{10}(f) + 49 \quad [\text{dB}] \quad . \quad (23)$$

where L represents the attenuation in dB, D is the distance in km and f is the carrier frequency equal to 2 GHz for WCDMA.

Path loss (23) is typically adopted for *Non Line-Of-Sight* (NLOS) propagation conditions between transmitter and receiver (worst-case).

The presence of big obstacles in the path between transmitters and receivers generates phases during which the signal is strongly attenuated, according to the *shadowing phenomenon*. In general, the signal slowly alternates between situations where it is slightly affected and intervals where its strength is strongly attenuated. The shadowing time variation is due to the MT motion relatively to the BS: since shadowing is produced by very big obstacles, we may understand that shadowing phenomena slowly vary with time. The shadowing attenuation is modeled through a log-normal variable with a certain degree of time correlation. Hence, the previous path loss model is typically coupled with a log-normally distributed shadow fading (standard deviation equal to 10 dB for outdoor users and to 12 dB for indoor users).

In urban scenarios there is another typical situation, when an MT turns at a street intersection: the signal strength is suddenly attenuated by about 20 dB (*corner effect*). This is the typical case of a Manhattan-like scenario where we have:

- Unobstructed LOS propagation conditions characterized by a direct path between the MT and the BS and

- Non-LOS propagation conditions due to the presence of obstacles, corners and so on.

A very simple way to characterize time-varying propagation conditions is given by the Gilbert-Elliott GOOD-BAD channel model described in [123],[124]. In the GOOD (BAD) state high (low) *Signal-to-Noise Ratio* (SNR) strength is experienced. The sojourn times in these two states are exponentially distributed with suitable mean values (Markov model). The two states are characterized by different bit error-rates (or packet error rates). This simplified model is based on the statistics of the SNR level crossing a given threshold in a non-frequency selective Rayleigh fading channel. Such model has been widely adopted in the literature to evaluate the performance of both cellular networks WLANs with microcellular coverage.

Let us denote:

P_g : probability of the GOOD state

P_b : probability of the BAD state

λ_{gb}: the mean transition rate from the GOOD state to the BAD state

λ_{bg} : the mean transition rate from the BAD state to the GOOD state.

The state probabilities of the Markov chain can be derived by means of the flow balance condition and the normalization condition:

$$\begin{cases} \lambda_{gb} P_g = \lambda_{bg} P_b \\ P_g + P_b = 1 \end{cases} . \tag{24}$$

Solving (24), we obtain:

$$P_g = \frac{\lambda_{bg}}{\lambda_{gb} + \lambda_{bg}} \quad \text{and} \quad P_b = \frac{\lambda_{gb}}{\lambda_{gb} + \lambda_{bg}} . \tag{25}$$

Let T_g denote the average sojourn time in the GOOD state (mean value $1/\lambda_{gb}$) and let T_b denote the average sojourn time in the *BAD* state (mean value $1/\lambda_{bg}$). The following laws have been determined:

$$T_b = \frac{e^{\eta} - 1}{f_D \sqrt{2\pi\eta}} \quad \text{and} \quad T_g = \frac{1}{f_D \sqrt{2\pi\eta}} \tag{26}$$

where:

f_D :   Doppler frequency shift, obtained as $f_D = Vf_p/c$, where V is the average MT speed in m/s; c is the light speed equal to 3×10^8 m/s; f_p is the carrier frequency in Hz.

η :   Dimensionless ratio between the threshold SNR value, τ_t, which discriminates between the GOOD state and the BAD state and the average SNR value experienced by a user, τ_0:

$$\eta = \frac{\tau_t}{\tau_0} \ . \tag{27}$$

Parameter τ_t depends on the mobile environment, the transmission techniques, the receiver implementation and the required signal quality; whereas τ_0 depends on the mobile environment, user distance, transmitted power, antenna characteristics and power control scheme.

It is important to note that the GOOD-BAD channel model is related to the user mobility by means of the mean user speed V that characterizes the Doppler frequency shift term f_D in T_b and T_g. In particular, if V increases, both T_b and T_g decrease, thus having more time-varying channel conditions. Typical V values range from 3 to 40 km/h in urban microcellular environments.

Let us denote with PER_g the *Packet Error Rate* in the GOOD state and with PER_b the *Packet Error Rate* in the BAD state. Hence, the mean packet error rate, PER, for a continuous packet traffic flow is:

$$PER = PER_g P_g + PER_b P_b \ . \tag{28}$$

GOOD-BAD channel models can be considered also in CDMA systems to model *outage phenomena* due to a sudden increase of the relative interference levels coming from simultaneous transmissions.

The GOOD-BAD model can be simplified to evaluate the outcome of packet transmissions according to the approach described below [125]. The product between the Doppler frequency shift f_D and the packet transmission time T_{pkt} determines the degree of correlation for the errors introduced on packet transmissions by the channel. If $f_D T_{pkt} > 0.2$ errors are packet-by-packet independent, with sufficient accuracy: a packet error occurs according to a given PER probability value obtained by sampling the GOOD-BAD process at one instant during the packet transmission. If $f_D T_{pkt} < 0.2$, packet errors are correlated and this simplified scheme cannot be adopted.

More refined channel models (suitable for layer 1 simulations) are beyond the scope of this book.

Chapter 3: RRM in GPRS

The success of present (2G and 2.5G) and future (3G) mobile networks depends on the provision of particularly attractive services to users. We refer here to the GSM system, where GPRS is used to access the Internet [22],[25],[126]. This Chapter deals with a detailed characterization of GPRS layer 2 along with a performance evaluation in the presence of Web interactive traffics.

3.1 Description of layer 2 protocols of GPRS

The *Logical Link Control* (LLC) layer is the first of three sub-layers that compose layer 2 of the GPRS air interface. LLC provides services to network protocol layers, as follows: it formats higher layer data units in LLC frames and delivers data to higher layers in the correct sequence. LLC is also responsible of information ciphering between the MT and the SGSN node.

The *Radio Link Control* (RLC) layer is the second sub-layer of OSI level 2 in the GPRS air interface protocol stack [25]. It is logically positioned between MAC and LLC. The principal RLC tasks are: (*i*) segmentation of LLC data and signaling frames (*Segment Data Units*, SDUs), into smaller standard fixed-length *Packet Data Units* (RLC/MAC-PDUs), also named *radio blocks*, and, vice versa; (*ii*) the re-assembly of RLC/MAC packets to obtain the original LLC frame. The LLC SDU maximum length is 1560 bytes with LLC headers, data and tail (*Frame Check Sequence*). The number of radio blocks needed to transmit an LLC SDU depends on the coding scheme used for transmissions. RLC is in charge of block numeration and manages two different transmission types with the MTs: "acknowledged" (ACK) and "unacknowledged". In ACK modality, the received blocks are sequentially reassembled in the final LLC SDU. While, in the second case, blocks are passed to the higher layers in the exact order that they are received; in this last case, higher layers will manage any data incoherence. Finally, RLC provides general control operations (identifiers of different data flows, and error managing) and synchronization functions (timers and counters for the different ACK and NACK connections) for the transmission of data.

The *Medium Access Control* (MAC) layer of GPRS [25] handles many important functions, such as the allocation of radio resources, the synchronization of the MT attempts to access the network (uplink), the forwarding of calls to MTs (downlink) and the optimization of the radio resource allocation to data and signaling channels depending on traffic conditions. The MAC layer is an interface towards the physical medium. For example, MAC has to deal with power control problems and with cell reselections due to the user mobility. The MAC layer, on the other hand, interfacing with RLC layer, has to provide numeration, ordering, and identification of each packet that it manages in order to re-build the original frame in the higher layers. This implies the management of a great number of counters and identifiers for each data flow. However, the main role of MAC is the management of the procedures that allow more MTs to share the same radio resources (i.e., slots). In fact, the MAC layer, with a dynamic management of radio resources, permits that an MT uses more slots of the same carrier so as to increase the data transfer speed with respect to the classical GSM system. MAC procedures define the creation of a TBF for the transfer of user (or signaling) data with the network. The MAC protocol also handles the TFIs to identify the different data flows on the physic medium.

Layer 2 resources are the GPRS logical channels that are used for data traffic (PDTCH), message broadcast (PBCCH), common signaling (PCCCH), traffic control (PACCH) and access network procedures (PRACH).

Each time an MT has to transmit data, a new access procedure has to be performed, because there is not a circuit dedicated to the user with GPRS. The amount of assigned radio resources can vary at each new connection between the network and an MT, and results a complex function of the different transmission types and traffic conditions.

3.2   Medium access modes

For each new TBF generated between the network and an MT, the GPRS system decides a specific radio resource (= block) allocation scheme, i.e., the method for transmitting (or receiving) different user data. In fact, the MAC layer decides when any single user has to transmit or receive data on the assigned time-slots. The system can assign to the MT one of three possible medium access modes, either in uplink or downlink [126]. In the following description, for the sake of simplicity, we will refer to uplink transmissions.

Fixed Slot Allocation: During the first step of the access procedure, radio resources are statically assigned: the network responds to the MT (that needs resources to transmit) with a particular message that specifies which time slots will be used by the MT. A map of the allocated slots ("ALLOCATION BITMAP") is used for this purpose.

Dynamic Slot Allocation: The network uses a 3-bit identifier USF (*Uplink State Flag*) in the header of the radio blocks that the network transmits in downlink to trigger the uplink transmission of each MT radio block. Each involved MT has to read and to decode the header of every block that the network sends in downlink to seek for its USF value: when the MT detects its USF in one or more assigned time slots, it realizes that it is the time to transmit its radio blocks in the corresponding time slots.

Extended Dynamic Slot Allocation: This resource allocation mode is used to simplify the MT transmission procedures. It is quite similar to the dynamic slot allocation mode, because it uses the USF flag, but with a different meaning: once the MT receives its USF on one of the assigned time slot, "X", $X \in [1, 8]$, it has to transmit its data on the corresponding time slot X and also in *all the following assigned* time slots. For example, let us consider an MT that is allowed to transmit in slots 2, 3, 5, and 7: if it receives the assigned USF from the network in time slot 2, it can transmit in time slots 2, 3, 5, and 7; while, if the MT receives the USF in time slot 5, it is allowed to transmit only in time slots 5 and 7.

3.3 Terminal states and transfer modes

An MT can be in two different modes: *activity*, when it is involved in a data transfer; *inactivity*, when it does not exchange data with the network. The procedures used in these two different states to start transmissions are deeply different.

Idle Mode: When the MT is not involved in data traffic, it is in "Idle Mode". In this state there is no traffic flow (TBF) between the MT and the network. The MT simply listens to the *Broadcast Channel* (PBCCH) of the cell and to some *Common system Control CHannels* (PCCCH) that can trigger new calls originated from the network (*Paging Channels*) [126]. The MT leaves this state only when it receives a call from the network (downlink) or when an access procedure has to be performed for uplink transmissions.

Packet Transfer Mode: When a data flow exists (in uplink or downlink) between the MT and the network (i.e., a TBF), the MT is in a "Packet Transfer Mode". In this state, radio resources (i.e., one or more physical channels) are assigned for point-to-point data transfer (that can be also bi-directional -uplink and downlink). In this transfer mode, two different transmission types are supported:

− **ACKNOWLEDGED transfer type:** after having received a fixed number of blocks an ACK/NACK message (uplink/downlink) is sent to the transmitter to notify the outcome of the transmission. In case of failure, a selective repeat retransmission is performed.

− **UNACKNOWLEDGED transfer type:** in this mode, the receiver still sends ACK/NACK as in the acknowledged mode, but these messages are only used to check the connection quality and to adapt the coding scheme to the radio conditions so as to guarantee a given error probability.

3.4   Access techniques

The MAC protocol is in charge of managing the procedures that allow an MT to start a communication with the network (uplink) and, vice versa, to receive traffic from the network (downlink). The access procedures will be explained in their main characteristics for the uplink direction (since downlink access methods are similar, once the initial paging phase has been performed).

3.4.1   *P*-persistent access procedure

Let us assume that an MT has already an active PDP context with the GPRS network. When the MT has an LLC frame to transmit, it has to open a communication context with the BTS by means of a TBF establishing procedure. Such TBF connection is requested by the MT with the PACKET CHANNEL REQUEST (PCR) message sent by means of a random access of the Slotted Aloha type on the *Pysical Random Access Channel* (PRACH). Since many MTs can simultaneously try to access the network, the access procedure is random on the PRACH channel. A settable number of blocks per multi-frame can be destined to PRACH on a slot per cell that represents the so-called *Master PDCH* (MPDCH). Since the PCR message occupies a single slot, a PRACH block can accommodate up to four PCR messages.

Collisions are resolved by GPRS with a special *P*-persistent access protocol. The MT acquires all the information and parameters necessary to define the *P*-persistent access procedure by listening to the PBCCH control channel that specifies:

-   The maximum number M of retransmissions that the MT can perform.

-   The persistence level P, with a value determined by the network; P can have four different values related to the four priority classes {1, 2, 3, 4}, named RLC/MAC *Radio Priorities* (RP). The default P value is 0.

-   Two values S and T that indicate to the MT the waiting time for the next attempt.

The MAC layer uses a certain number of timers related to the access procedure deadline and counters to manage the synchronization of the different operations.

The detailed description of the access procedure is provided in the following sub-Section 3.4.5.

3.4.2   One- and two-phase access procedures

In order to reduce the risk of collisions among MTs that simultaneously try to access the network on the same PRACH, the PCR message sent by MTs is quite short. If the MT has to negotiate with the system a more detailed resource request, it can ask to the network the opportunity for sending a second request message, after the first short one. This is the so-called "two-phase access" procedure. In fact, once the network has correctly received (without collisions) the first request message, the network sends a message to assign a resource on a physical channel, where the MT can send the second (and longer) request message. Then, the network responds with a special message (PACKET UPLINK ASSIGNMENT) that specifies the radio resources allocated to each MT for the transmission of data. In the "acknowledged" transfer modality, the access procedure (with either single or two phases) is terminated when the MT receives from the network an ACK/NACK related to the first transmitted packet.

3.4.3   Queuing and polling procedures

If the network is unable to assign the requested resources to the MT, no response is sent back after a PCR message is correctly received on PRACH. The MT not receiving any message from the network retries the access procedure according to the GPRS *P*-persistent scheme. Anyway, in order to avoid MTs sending too many access messages, the network can acknowledge the MT with a message named PACKET QUEUEING NOTIFICATION, so that the MT does not send other access requests. This message notifies that radio resources will be assigned as soon as available. Meanwhile, the network can interrogate the MT to verify whether it is still waiting for resources to be assigned. In this case, the network sends a PACKET POLLING REQUEST message, which notifies the MT that there are not yet available resources. The MT, in turn, responds with a simple PACKET CONTROL ACK message to confirm that is still waiting for resources. Queuing and polling procedures terminate when the network assigns a resource to the MT or when the MT has waited for more than 5 s after the queuing message.

3.4.4   Paging procedure

The network can page an MT to start a downlink data transfer using a dedicated logical channel (paging channel) always listened by all the MTs in the cell. The procedures that create a downlink connection between the network and an MT are quite similar to those used in uplink. In fact, after the first paging message from the network, the MT performs an access procedure to have assigned a channel for sending its response. The network then assigns a radio resource that the MT uses for its reply to the paging message. Moreover, the network sends a specific control message that indicates the beginning of the downlink data transfer.

3.4.5   A detailed example of a one-phase access procedure

We describe below an access procedure with queuing and polling for dynamic slot allocation and the acknowledged transmission type. The following description refers to the signaling diagram shown in Fig. 7.

An MT needing a connection with the network, first listens to the system information sent in PBCCH to acquire the values of all the parameters necessary for the P-persistent access procedure (sub-Section 3.4.1). Then, the P-persistent access procedure is performed as described below [25].

The first attempt to send a PCR message can be initiated at the first possible TDMA frame containing the PRACH. For each attempt, the MT extracts a random value R: the MT is allowed to send the PCR message only if R is greater than or equal to P. After a request is issued, the MT waits a time, which depends on S and T. If the MT does not receive the PACKET UPLINK ASSIGNMENT in this time, a new attempt is tried, if it is still allowed, otherwise a failure is notified to the higher layer. After consuming $M+1$ attempts, the MT waits a time, which depends on S and T, and if it does not receive the PACKET UPLINK ASSIGNMENT, a packet failure is notified to the higher layer. The number of retransmissions per access attempt is limited to 1, 2, 4, and 7 for the RLC/MAC RP 1, 2, 3 and 4, respectively. The MAC layer also notifies to the higher layer the failure of the access procedure also if more than 5 s (timer T3186) last from the first access message without any reply from the network.

Note that if the network has correctly received a PCR message from the MT, but there is not enough available radio resources, the network can notify the MT to stop sending its requests with a PACKET QUEUEING NOTIFICATION message that informs the MT to wait at most 5 s (timer T3162) for a valid resource assignment. If timer T3162 expires without a valid resource assignment, the MAC layer declares that the access procedure is failed. When, the MT is waiting for a valid uplink assignment, the network can ask the MT whether it is still waiting for resources in the cell by means of a polling message, named PACKET POLLING REQUEST. The MT simply acknowledges this message with a PACKET CONTROL ACK. If a dynamic slot allocation is chosen for the data transfer, as soon as the network has enough radio resources, a PACKET UPLINK ASSIGNMENT message is sent to the MT. This message specifies both the numbers of time used slots and a 3-bit USF to label the transfer. Each time the network will send the specified USF value in its downlink block headers, the MT will transmit its data on the time slots where the USF has been received.

Finally, if the connection type is "acknowledged", the network periodically sends an ACK/NACK message to the MT, for a selective retransmission of erroneous blocks. The one-phase access procedure ends (i.e., the contention is resolved) when an ACK/NACK message is received from the network for the first sent packet.

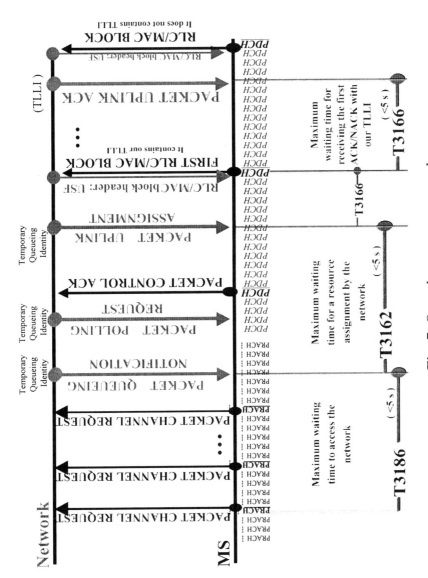

Fig. 7: One-phase access procedure.

3.5   GPRS performance evaluation

On the air interface, physical GSM channels (i.e., slots) destined to GPRS are denoted as *Packet Data CHannel* (PDCH) and have a 52-frame organization. The basic transmission unit on a PDCH is called *radio block*: a slot in four consecutive frames is utilized to transmit a radio block. Every 13-th frame, the slot of the PDCH is not used to transmit data, so that there are 12 radio blocks per multi-frame. A radio block contains 456 bits, but the number of information bits depends on the coding scheme: for coding schemes CS-1, CS-2, CS-3 and CS-4, a block conveys about 22, 33, 38 and 53 information bytes and correspondingly, 9.06, 13.4, 15.6 and 21.4 kbit/s are achieved for the use of one slot (PDCH) per frame.

The following performance evaluation focuses on downlink, the system bottleneck for Internet browsing. A multi-slot MT can be assigned up to eight slots (practical implementation aspects reduce this theoretical limit to four slots) per frame. We refer here to MTs that have negotiated suitable QoS profiles with the SGSN through the activation of PDP contexts and that have established TBFs for the downlink transfer of data.

As explained in Chapter 1 of Part II, for downlink transmissions, we envisage a queue at the BS for each traffic class: conversational (telephony, transactional services), streaming (video streaming, ftp), interactive (Web browsing) and background (e-mail traffic) [26].

In this study, we consider only two different traffics (hence, two queues): conversational class (here used to support transactional services for mobile users) and interactive class for Web surfing traffics. The conversational traffic is assumed to have a preemptive resume priority with respect to the other traffic class. When there is no conversational traffic to be managed, radio blocks are assigned to serve the interactive class queue. Slots are dynamically assigned to MTs according to their TBFs.

Conversational traffic due to transactional applications is assumed to produce a Bernoulli block arrival process with r Erlang/PDCH. As for interactive traffic, the Web model shown in Chapter 2 of Part II has been considered with $q = 1$, that corresponds to a mean bit-rate of 5.83 kbit/s. Let us assume that M_w simultaneously Web browsing users share the resources of a given GPRS carrier. We evaluate the performance experienced by these users in terms of the mean datagram transmission delay, $E[t_{delay}]$. A simulator has been built to characterize the assignment of the transmission PDCH resources to competing traffics at OSI layer 2.

Fig. 8 shows $E[t_{delay}]$ as a function of the number of competing downlink interactive traffic flows (users) for the GPRS CS-2 case considering n PDCHs per frame ($n = 2$, 3 and 4, the maximum value for present GPRS implementations); *dashed lines (continuous lines)* are for $nr = 0.1$ Erlang ($nr = 0$ Erlang). FIFO scheduling has been assumed for the transmission of datagrams. We may note that $E[t_{delay}]$ increases with the number of users and decreases with the number of PDCHs destined to GPRS. Note that the presence of the conversational traffic causes an increase of $E[t_{delay}]$. This graph can be useful for dimensioning the number of PDCH resources as a function of requirements on $E[t_{delay}]$.

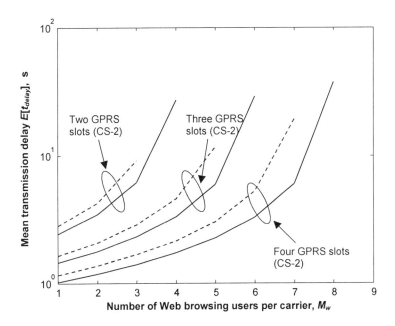

Fig. 8: $E[t_{delay}]$ as a function of the number of competing downlink interactive traffic flows for GPRS CS-2 coding scheme and n PDCHs

($n = 2, 3, 4$). *Continuous lines* (*dashed lines*) are for cases with no conversational traffics (conversational traffic load equal to 0.1 Erlang).

Fig. 9 presents the $E[t_{delay}]$ behavior as a function of the number of downlink interactive traffic flows, M_w, in a case with 2 PDCHs and different GPRS coding schemes. The curves stop for the maximum M value beyond which the traffic load cannot be supported by the GPRS downlink transmission queue for interactive traffics (instability). It is interesting to note that $E[t_{delay}]$ significantly decreases from CS-1 to CS-4 and that the maximum M_w value with CS-4 is much greater than with CS-1.

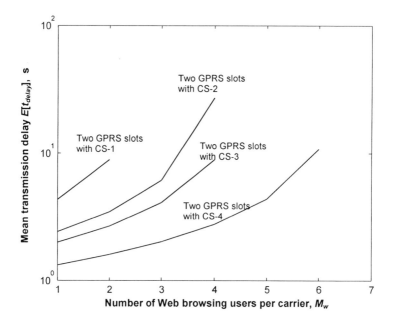

Fig. 9: $E[t_{delay}]$ as a function of the number of competing downlink interactive traffic flows with 2 PDCHs for different coding schemes and no conversational traffic.

We consider now a *Round Robin* (RR) technique applied to the downlink queue to manage transmissions on a given GPRS carrier. As shown in the Chapter 1 of Part II, RR is a particularly efficient solution in the presence of heavy-tailed message length distributions.

Two different RR schemes can be considered for the management of the datagrams coming from different users in the interactive transmission queue (see Fig. 10):

- *Datagram-based RR*: one entire datagram is transmitted per user per cycle;

- *Block-based RR*: one block of a given datagram is transmitted per user per cycle.

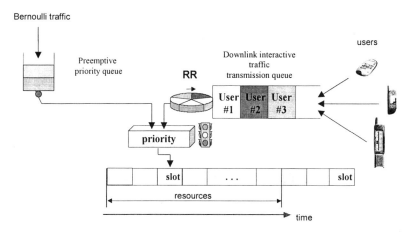

Fig. 10: RR service policy for downlink.

Fig. 11 compares the *cumulative distribution functions* (cdfs) of the message delay obtained from simulations for the two RR schemes with 1 PDCH per frame, $M_d = 2$, $r = 0.1$ Erlang and Poisson arrivals of geometrically distributed messages (see the data traffic model in Chapter 2 of Part II) with mean arrival rate of 1.52 msg/s and mean message length of 481 bytes/msg. We note that, even if the block-based RR scheme guarantees a maximum delay value for block transmissions (thus supporting minimum bit-rate values per MT), the datagram/message-based RR scheme achieves much better results for 90-th and 95-th percentiles of the message delay, thus improving the behavior of applications and the QoS perceived by users.

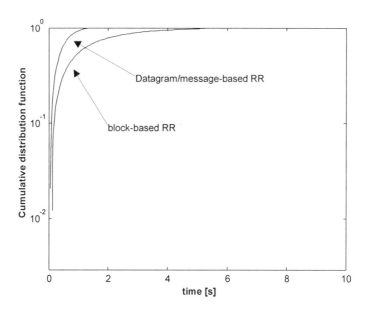

Fig. 11: Cdfs of the message transmission delay in the GPRS CS-4 case with 1 PDCH/frame, $r = 0.1$ Erlang, $M_d = 2$.

Chapter 4: RRM in WCDMA

We refer here to the packet-switched management of bursty traffics (i.e., video and Web traffics)on WCDMA resources [127]. We foresee here an RRM scheme based on a scheduler that, operating at the Node B level (or at the RNC level), dynamically decides on a frame basis, the transmission of pending messages from the different MTs (uplink). Resource allocation is accomplished by defining processing gain (on a message/datagram basis) and transmission power (on a frame basis).

Since we refer here to uplink transmissions, we assume that a video or a Web traffic source makes a transmission request for each new video frame or for each new datagram to be transmitted to the Node B. These transmissions occurs through the RACH channel for MTs not engaged in communications; otherwise, we assume that an associated control channel is used to notify new requests (piggybacking scheme). Finally, we assume that each request contains the length of the message to be transmitted.

We start this study by considering the *Dynamic Resource Scheduling* (DRS) technique proposed by Ö. Gürbüz and H. Owen in [128]-[131]. This is an optimized power management scheme for mobile users with different traffic classes under specific constraints in terms of energy per bit-to-noise and interference power spectral density, $E_b/I_{0,tot}$ [132]. We have integrated the DRS technique with the selection of the processing gain on a message basis, according to the rate-matching scheme of the WCDMA air interface. Moreover, we have also introduced a request service priority order among the different traffic classes to support efficiently real-time video and Web bursty traffics.

In the following derivations, we consider MTs that randomly move within in a cell. The user mobility model and the channel propagation conditions will be described later in this Chapter. Let us examine the algorithm to determine the power levels used by MTs to transmit to the Node B according to $E_b/I_{0,tot}$ requirements.

In particular, we may write the following $E_b/I_{0,tot}$ equation with related QoS constraint for the *i*-th traffic source or MT ($i = 1, 2, \dots N$):

$$\frac{E_b}{I_{0,tot}}\bigg|_i = \frac{PG_i h_i P_i}{\displaystyle\sum_{j\neq i, j=1}^{N} h_j P_j + \eta_o W} \geq \gamma_i \tag{29}$$

where:

- The summation at the denominator for $j \neq i$ takes account of the intra-cell interference;

- N represents the number of transmission requests to be managed in the current frame (if room);

- γ_i is the minimum requested $E_b/I_{0,tot}$ value (QoS requirement) for the i-th user;

- PG_i is the processing gain of the i-th user;

- h_i is the path gain due to the distance of the i-th user from the Node B;

- W is the total spreading bandwidth;

- P_i is the transmitted power for the i-th user in the current frame;

- η_o is the power spectral density due to both background noise and inter-cell interference (both modeled as white Gaussian noise with a suitable variance).

In this study we assume that the Node B adopts a single-user receiver so that intra-cell interference cannot be avoided. Note that extra-cell interference can be considered as a fraction of the intra-cell interference by means of a white Gaussian noise whose variance depends on both the power control mechanism and the channel propagation conditions. Intra- and inter-cell interference can be characterized according to the *Standard Gaussian Approximation*, due to the presence of many equal interfering contributions (*central limit theorem*).

For each frame, the scheduler performs a fast resource allocation scheme by assigning power levels to the different requests, each identified by h_i and PG_i values. Note that the Node B knows the h_i value according to the power control scheme (an ideal scheme is assumed in this study) and determines the PG_i value on the basis of the rate-matching algorithm adopted for each transmission request.

We define a resource (power) allocation scheme to the different requests considering the rigid fulfillment of QoS requirements for the different transmissions:

$$\frac{PG_i h_i P_i}{\sum_{j \neq i, j=1}^{N} h_j P_j + \eta_o W} = \gamma_i , \quad \forall i = 1, 2, \dots N .$$ (30)

This equality can be re-written as:

$$PG_i P_i = \frac{1}{h_i} \left[\left(\sum_{j=1}^{N} h_j P_j + \eta_o W \right) - h_i P_i \right] \gamma_i$$ (31)

or equivalently

$$P_i \left(PG_i + \gamma_i \right) = \frac{1}{h_i} \left(\sum_{j=1}^{N} h_j P_j + \eta_o W \right) \gamma_i .$$ (32)

Let us introduce the following quantity:

$$g_i = \frac{\gamma_i}{\gamma_i + PG_i} .$$ (33)

that represents a *power index* to weight the amount of resources that need to be assigned to a given transmission requests. From (32) and (33), we have:

$$h_i P_i = g_i \left(\sum_{j=1}^{N} h_j P_j + \eta_o W \right) , \quad \forall i = 1, 2, \dots N .$$ (34)

We sum the expressions in (34) for each *i* value; we have:

$$\sum_{i=1}^{N} h_i P_i = \sum_{i=1}^{N} g_i \times \left(\sum_{j=1}^{N} h_j P_j + \eta_o W \right) .$$ (35)

Through some algebraic manipulations, (35) can be written as:

$$\sum_{i=1}^{N} h_i P_i = \frac{\eta_o W \sum_{i=1}^{N} g_i}{\left(1 - \sum_{j=1}^{N} g_j\right)} . \tag{36}$$

Hence, by substituting (36) in (34), we can solve P_i as:

$$P_i = \frac{g_i \eta_o W}{h_i \left(1 - \sum_{j=1}^{N} g_j\right)} , \quad \forall i = 1, 2, ..., N . \tag{37}$$

The previous formula defines the optimum power allocation method to fulfill QoS requirements. We introduce a priority order according to which requests are served in the frame. We progressively admit new requests to be served and compute the related power levels; accordingly, we also update the N value: $N \leftarrow N + 1$. For each new admitted request we have to verify whether the following condition is fulfilled:

$$\sum_{j=1}^{N} g_j < 1 . \tag{38}$$

If this *Connection Admission Control* (CAC) criterion is not fulfilled, the new request cannot be served in the current frame.

Practically, we simply evaluate the CAC tests by progressively allocating new requests according to the priority order. Finally, when all the requests have been admitted or the CAC test has stopped new allocations, the power levels are computed according to (37).

In real implementations, each MT has a maximum transmit power level, P_{max}. Once power levels P_i have been assigned according to (37), we may verify whether they exceed P_{max}. In these circumstances, the MT transmits at the following power level:

$$\hat{P}_i = \min(P_i, P_{max}) . \tag{39}$$

We assume that an *outage event* occurs (the transmitted packet is lost since the required $E_b/I_{0,tot}$ cannot be achieved) if $P_i > P_{max}$.

In general, the P_{max} value depends on the MT type.

4.1   Adopted models

In the UMTS 30.03 document [53], three different test environments have been proposed, each of them characterized by suitable mobility and propagation conditions: *vehicular, indoor, Outdoor-to-Indoor and Pedestrian* test environments. Let us refer here to the third case, proposed for urban (Manhattan-like) scenarios. The path loss model depends on the fourth power of the distance between the MT and the Node B, as described in Chapter 2 of Part II. For this study, h_i coincides with the L value in (23) that depends on the MT distance from the Node B.

As for the MT mobility, we consider users that move within a tagged cell; at regular time intervals their position is updated. Correspondingly, h_i varies and it is estimated at the Node B with an ideal closed-loop power control scheme.

The Node B performs RRM on the basis of the estimated h_i value. The resource allocation is kept fixed for all the frame duration. Since we consider slow moving users, we can neglect the impact on the QoS due to h_i variations in a frame interval of 10 ms (otherwise, suitable guard margins should be taken on the target γ_i values).

For the sake of simplicity, we assume that cells are circularly shaped (cell radius of 500 m) and that MTs move according to straight trajectories. When an MT reaches cell boundaries, its motion direction is reflected towards the cell in a way to have both a uniform distribution of users per cell and a constant number of users per cell[12]. At each motion change, the MT speed is regenerated according to a uniform distribution from 0 to 30 km/h.

---

[12] We do not consider handoffs that would entail the introduction of complex power control mechanisms for the cells. For more details related to the management of handoffs, the interested reader may refer to [133].

As for Web traffic sources, we have used the model shown in Chapter 2 of Part II with $q = 1$ [53]. For video sources, we have adopted the D-MAP model shown in Chapter 2 of Part II (evolving on a frame $= 10$ ms basis) with $M = 5$ minisources, mean bit-rate $\mu = 144$ kbit/s, standard deviation $\sigma = 101$ kbit/s, autovariance parameter $a = 3.9$ s^{-1} (correspondingly, $p = 37$ frames, $q = 92$ frames, $\psi_v = 0.286$, $A = 100$ kbit/s, maximum bit-rate of 500 kbit/s) [114].

4.2   Detailed description of the proposed RRM scheme

For each datagram to be transmitted by an MT, we propose that the Node B associates the generation bit-rate, obtained as the ratio between the datagram length and the estimated interarrival time. This mechanism allows to control in some way the service of datagrams (*elastic capacity concept*). If fact, if datagrams arrive sufficiently spaced, a low bit-rate is assigned; otherwise, if datagrams arrive in bursts, they are served with a higher bit-rate to reduce queuing times.

As for video sources, a similar approach is adopted: we assume that the video source updates the Node B at regular intervals ($= 40$ ms) with its transmission needs; this request contains the number of bits generated in the interval (here referred to as a *picture*). Hence, the corresponding generation bit-rate is computed as the ratio between the number of bits and the time interval.

These bit-rate values (one for each request) are used by the rate-matching algorithm to map them into the appropriate bit-rate supported by the WCDMA air interface with related processing gain values PG_i. The available net bit-rate values are {15, 30, 60, 120, 240, 480, 960} kbit/s with related processing gain values {256, 128, 64, 32, 16, 8, 4}. Note that the effective bit-rate values are obtained by considering that the above values are reduced due to both a channel coding with rate 1/2 and a protocol overhead due to higher protocol layers (here considered of 5%).

The i-th transmission request received at the Node B is characterized by h_i, PG_i (evaluated as explained above and, hence, the bit-rate R_i) and the number of bits, L_i. Video traffic transmission requests are served before Web traffic requests. Within each traffic class the Node B uses a priority order that depends on the service time (i.e., $T_i = R_i L_i$) associated with each request. In particular, different prioritization schemes have been compared [134].

- *Longest Processing Time* (LPT) policy: the highest priority request is that with the highest T_i value.

- *Shortest Processing Time* (STP) policy: the highest priority request is that with the lowest T_i value.

- *Longest Global Time* (LGT) policy is a refinement of the LPT scheme: the highest priority is given to the request with the highest $T_i + D_i$ value, where D_i denotes the waiting time experienced by the *i*-th request before being served.

The LPT criterion selects the request that entails a longer service time; this request may correspond to exceptionally long messages or (more commonly) to messages requiring low R_i values (high PG_i values) and, hence, low transmission power values, thus causing low interference. Whereas, the SPT criterion decides to serve the request that yields a higher interference level, but for a shorter time (high R_i value, low PG_i value and high power level).

Resources are progressively allocated to requests on the basis of their priority, defining the power levels according to (37) and (39) and fulfilling the CAC constraint (38). Progressively serving more requests, the $\sum_{j=1}^{N} g_j$ value increases as well. Correspondingly, we have also an increase in the power levels assigned to the MTs. If we allow too much requests admitted in the system (even with the CAC condition fulfilled), we may have that some power levels exceed the allowed maximum, so that the corresponding transmissions experience outage. This drawback could be only avoided by making more complex controls, not addressed here. In fact, we simply consider that the allocation process stops as soon as the inclusion of a new request causes the violation of the CAC condition (38). The requests not served in the current frame are reconsidered in the next one (*removal*).

The above RRM scheme could also support an adaptive selection of QoS levels in terms of γ_i values. In particular, we could reduce the γ_i requirements in the presence of heavy traffics. The study of these aspects is beyond the scope of this investigation.

4.3   Simulation results

We show here simulation results obtained by implementing the previous RRM schemes, traffic sources, user mobility and path loss models. Moreover, the following data have been adopted for numerical evaluations: $\eta_0 = -174$ dBm/Hz, $P_{max} = 250$ mW for data terminals and $P_{max} = 750$ mW for video terminals. As for γ_i values, they depend on many aspects, and, in particular on the traffic type and the radio mobile environment. We have adopted $\gamma_i = 3.2$ dB for Web traffics and $\gamma_i = 1.4$ dB for video traffics.

We discuss below the results obtained in terms of mean datagram transmission delay for Web traffic sources and mean delay to transmit a block of bits generated during a 40 ms period (i.e., a *picture*) for video sources. Figs. 12 and 13 show the mean datagram delay and the mean picture delay for the DRS scheme with different prioritization techniques (i.e., LPT, LGT and SPT) for individual traffic sources in a configuration with 20 Web users and 4 mobile video users per cells. We may note that all the video and all the Web sources experience almost the same QoS levels (*fairness*). Interestingly, the SPT scheme strongly outperforms all the other schemes in terms of delays. Hence, it is better to send first the request with the shortest transmission time. Analogous considerations can be obtained in terms of the outage probability that in these cases is around equal to 2% for all the techniques, with SPT allowing the best performance.

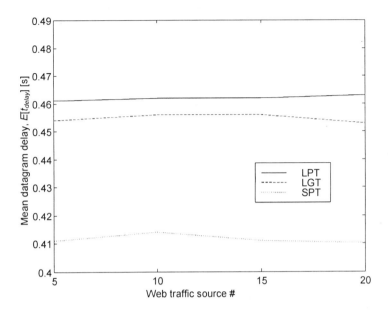

Fig. 12: Comparison of the prioritization schemes SPT, LGT and LPT in terms of mean datagram delay.

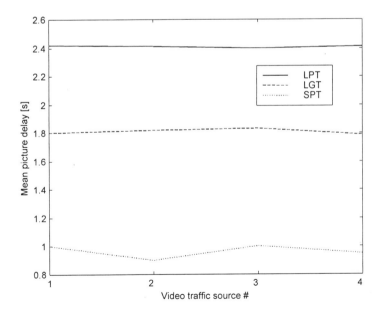

Fig. 13: Comparison of the prioritization schemes SPT, LGT and LPT in terms of mean picture delay.

Finally, in the case of the SPT technique that has shown the most promising results, we have evaluated the probability that a video picture cannot be served in time (thus it is dropped) in a configuration with 5 Web traffic sources and 4 video sources. The picture deadline has been here assumed equal to 100 ms. We have obtained a dropping probability of about 0.1% with a corresponding outage probability around 3%.

A possible refinement for these RRM schemes could be to avoid transmissions when the required power level exceeds the P_{max} value (outage event). This strategy is particularly useful for Web traffics that have not a deadline to be fulfilled and that can be delayed until more favorable channel conditions are experienced. The same approach is practically useless for real-time traffics, since delayed packets risk to be dropped due to deadline expiration.

Chapter 5: RRM in UTRA-TDD

Fig. 14 describes the general UMTS protocol architecture (for both FDD and TDD air interfaces) in terms of its entities: *User Equipment* (UE), UTRAN and *Core Network* (CN) [72]. The reference points relative to Uu (Radio Interface) and Iu (CN-UTRAN Interface) are also shown. The *Access Stratum* (AS) is a functional grouping of all the layers (in UE, Node B and RNC) related to the radio access network; AS boundaries are between the layers that are independent of the access technique and those that depend on it. AS offers services through the following *Service Access Points* (SAP) to the *Non-Access Stratum* (NAS):

- *General Control* (GC) SAPs,

- *Notification* (Nt) SAPs,

- *Dedicated Control* (DC) SAPs.

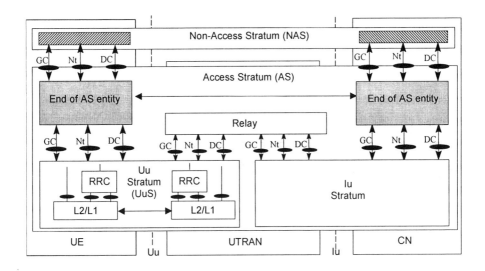

Fig. 14: Conceptual architecture of the UMTS protocols (SAPs are marked by circles).

The following description even if referred to the UTRA-TDD case is also valid (with few changes) for the WCDMA FDD case.

5.1   Radio interface protocol architecture: details

The UMTS radio interface is structured in three protocol levels [72],[76],[135]:

- *Physical layer* (L1)

- *Data link layer* (L2)

- *Network level* (L3).

Level 2 is divided into the following sub-layers: *Medium Access Control* (MAC), *Radio Link Control* (RLC), *Packet Data Convergence Protocol* (PDCP) and *Broadcast/Multicast Control* (BMC).

Level 3 and RLC are divided into *Control* (C-) and *User* (U-) planes.

Fig. 15 shows the radio interface protocol architecture. Every block represents a protocol. The SAP between MAC and the physical level provides the transport channels. SAPs between RLC and MAC provide the logical channels. The RLC level has three SAP types, one for every RLC operation mode: AM (*Acknowledged Mode*), TM (*Transparent Mode*), UM (*Unacknowledged Mode*).

The RLC sub-layer provides ARQ (*Automatic Repeat reQuest*) functionalities. RLC has no difference between control and user plane.

The transport service provided by layer 2 is called *radio bearer*. The radio services of the control plane, provided by RLC to RRC, are denoted as *signaling radio bearers*.

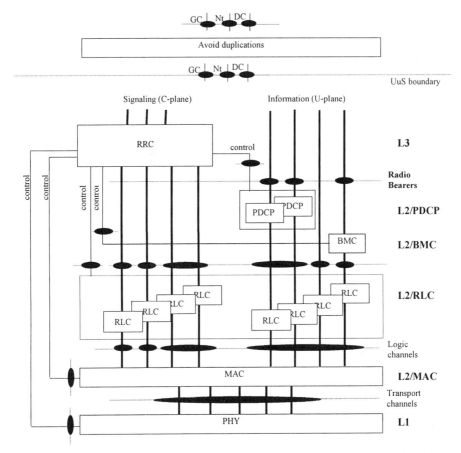

Fig. 15: Radio interface protocol architecture (SAPs are marked by circles). In this figure we can easily locale logic channels, transport channels and physical channels.

The following description is related to the TDD version of UMTS, as standardized by 3GPP starting from Release 4.

5.2 Transport and physical channels

The physical layer offers information transfer services to both MAC and higher layers. Such transport services are described on the basis of *how* and *in which format* data are transferred over the radio interface. Resources at this layer are named *transport channels* (clearly distinct from the classification of *what* is transported that refers to logical channels). *Transport CHannels* (TrCH) can be divided into two groups [67]:

- *Common transport channels*, where in-band signaling is used to identify UEs;

- *Dedicated transport channels*, where resources identified by code, time-slot and frequency are rigidly assigned to UEs.

There are six types of common transport channels in the TDD mode:

- BCH - *Broadcast CHannel* is a downlink transport channel, used to transmit system and cell-specific information.

- FACH - *Forward Access CHannel* is a downlink transport channel, used to carry control information to a UE when the system knows its position. FACH can also carry short user packets to the UE (sporadic data traffic).

- PCH - *Paging CHannel* is a downlink transport channel, used to carry control information to a UE in order to identify its position.

- RACH - *Random Access CHannel* is an uplink transport channel that is used to transport control information (i.e., access requests) from the UE. RACH can also carry short user packets (sporadic data traffic).

- USCH - *Uplink Shared CHannel* is an uplink transport channel shared by several users for dedicated control or data traffic (TDD mode only).

- DSCH - *Downlink Shared CHannel* is a downlink transport channel shared by several users for dedicated control or data traffic.

As for dedicated transport channels, *Dedicated CHannel* (DCH) is an up-/down-link transport channel that is used to carry control or user information between UTRAN and UE.

Transport channels are the interface to communicate between MAC and physical layer. Let us focus on some terminology on transport channels, as detailed below.

- *Transport Block* (TB): A transport block is the basic data unit exchanged between physical and MAC layers.

- *Transport Block Set* (TBS): A transport block set is a collection of transport blocks that are sent over a given transport channel.

- *Transmission Time Interval* (TTI): This is the inter-arrival time of TBS delivered from MAC to the physical layer. It is determined by the interleaving scheme in operation on the given transport channel.

The information managed at the transport channel level is organized in TBs (a TB typically corresponds to an RLC-PDU). The TTI time that can be an integer multiple of the frame length (from 10 ms to 80 ms) represents the interval according to which a TBS is forwarded from MAC to the physical channels after multiplexing and coding operations.

The rules for the delivery of data on transport channels are specified in the *Transport Format* (TF) that contains the following information:

Dynamic part:

- Transport Block Size
- Transport Block Set Size

Semi Static part:

- TTI
- Coding Type
- Coding Rate
- Rate Matching Parameter
- CRC length.

For example, a TF specification is as follows: dynamic part = {320 bits, 2 blocks}, semi static part = {40 ms, turbo coding, rate 1/3, 1.25, 16}.

Another important aspect associated with a transport channel is the *Transport Format Set* (TFS). A TFS is the set of TFs allowed on a given transport channel. Note that a TF is associated to each transport channel with a fixed (or slowly varying) rate. Whereas, a TFS is associated to each transport channel with a rapidly changing rate. The available TFs in the TFS are indexed by an integer number, called the *Transport Format Indicator* (TFI).

A variable rate DCH has one TF for each rate (i.e., a TFS); whereas, a fixed-rate DCH has only associated one TF.

Each UE can simultaneously have more active transport channels (each of them with specific TFs): the physical layer has to multiplex them on one or more physical channels. The data flux resulting from such operation of multiplexing and coding is named *Coded Composite Transport Channel* (CCTrCH) and may be transmitted on more physical channels [68].

Since, the physical layer can multiplex several transport channels together, each having its own TFS, it is necessary to specify the format of the composite channel after multiplexing, CCTrCH. This function is accomplished by specifying a vector consisting of an element from the TFS for each of the transport channels. This vector is called a *Transport Format Combination* (TFC). A TFC can be specified by giving a TFI value for each transport channel. The set of TFCs is indexed by a number called the *Transport Format Combination Indicator* (TFCI). The set of allowed TFCs is called the *Transport Format Combination Set* (TFCS).

5.2.1   Spreading for downlink and uplink physical channels

Downlink physical channels use PG = 16 [67],[69],[136]. Several parallel physical channels can be used to support higher data rates for the transmission with a user. These parallel physical channels will be sent by using different channelisation OVSF codes. Spreading operation with a single spreading factor 1 is possible for the downlink physical channels, but with very optimal channel conditions and very low multiple access interference levels.

The processing gain values that can be used for uplink physical channels range from 16 down to 1 [67],[69]. For every physical channel an individual minimum processing gain value, PG_{min}, is assigned from higher layer protocols. There are two options that are decided by UTRAN:

- The UE uses the processing gain PG value independently of the current TFC.

- The UE autonomously increases the PG value, according to the current TFC; the UE shall vary the code along the branch with the higher code numbering of the allowed OVSF sub-tree (i.e., greater PG value).

For uplink physical channels, single-code transmission (with PG values from 1 to 16) is better than multi-code one, because a smaller peak-to-average transmission power ratio is obtained, thus reducing battery consumption. Moreover, it enables the implementation of more efficient power amplifiers in the UEs.

For a multi-code uplink transmission, a UE shall use a maximum of two physical channels per time slot. These two parallel physical channels are transmitted by using distinct OVFS codes.

Three *burst types* are used for physical channels depending on the applications. All of them are formed by two fields for data symbols, a midamble and a *Guard Period* (GP), as shown in sub-Section 3.4.3 of Part I. The number of data symbols in a burst depends on the PG value and the burst type, as shown in the Table 2 [67].

Processing gain (PG)	**Burst Type 1**	**Burst Type 2**	**Burst Type 3**
1	1952	2208	1856
2	976	1104	928
4	488	552	464
8	244	276	232
16	122	138	116

Table 2: Number of data symbols for burst types 1, 2 and 3.

All burst types (1, 2 and 3) provide the possibility for the transmission of TFCI (Fig. 16) to specify the format adopted in the related bursts. The transmission of TFCI is negotiated at call setup and can be renegotiated during the call. A time slot can carry or not the TFCI (individually signaled). The TFCI is always inserted in the first time slot of a frame for every CCTrCH.

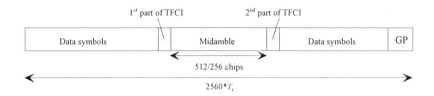

Fig. 16: Position of the TFCI code word in the burst (downlink).

Burst type 1 is suited for uplink if more than three UEs share one time slot (max 8). Burst type 2 can be used in downlink and also in uplink, if the bursts within a time slot are allocated to less than four UEs [136].

In the burst, the midamble may also denote the UE for a dynamic channel allocation. There are three different schemes of midamble allocation for physical channels:

- **Default midamble allocation**: the midamble for uplink or downlink is defined by layer 1 according to a mapping related to the assigned channelization code and burst type.

- **UE-specific midamble allocation**: a UE-specific midamble for uplink or downlink is explicitly assigned by higher layers.

- **Common midamble allocation**: the midamble for downlink is assigned by layer 1 depending on the number of channelization codes used in the considered downlink time slot.

If the midamble is not explicitly assigned and the use of a common midamble is not signaled by higher layers, the midamble is allocated by layer 1, according to the default scheme.

5.2.2 Multiplexing, channel coding and interleaving

The total number of basic physical channels (*Resource Units*, RUs) per frame on a carrier frequency is given by the maximum number of time slots (= 15) and the maximum number of CDMA codes per time slot (= 16). Data arrive to the coding/multiplexing unit in transport RLC blocks at regular TTI intervals that are transport channel specific and belong to the set {10 ms, 20 ms, 40 ms, 80 ms}. Coding/multiplexing involves the following steps:

- Add *Cyclic Redundancy Check* (CRC) to every transport block for error detection.

- Transport block concatenation/code block segmentation: all transport blocks in a TTI are serially concatenated.

- Channel coding. The channel coding schemes that can be applied to transport channels are shown in Table 3.

TrCH type	Coding scheme	Code rate
BCH	Convolutional	1/2
PCH	"	1/2
RACH	"	1/2
DCH, DSCH, FACH, USCH	"	1/3, 1/2
"	Turbo coding	1/3
"	No coding	

Table 3: Channel coding types.

5.3 MAC layer

The MAC protocol architecture can be described in terms of MAC entities, the most important of which are described below (see Fig. 17) [76].

- MAC-b is the MAC entity that manages the transport *Broadcast Channel* (BCH).

- MAC-c/sh is the MAC entity related to the following common transport channels: PCH, FACH, RACH, DSCH and USCH.

- MAC-d is the MAC entity that handles *Dedicated transport Channels* (DCH). The MAC-d entity has a link to the MAC-c/sh entity to transmit or receive data on transport channels that are handled by MAC-c/sh (uplink/downlink).

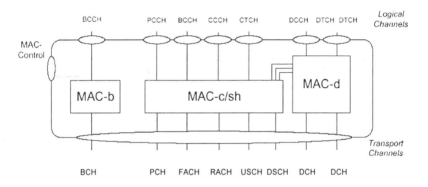

Fig. 17: MAC entities architecture.

These MAC sub-protocols have different tasks in UE and UTRAN, as described below.

MAC-c/sh entity - UE side

The MAC-c/sh entity has the following main functions from the UE side (Fig. 18):

- *Target Channel Type Field* (TCF) multiplexing ([13]) function for the handling of the TCF field in the MAC PDU header and related mapping between logical and transport channels (uplink: insertion; downlink: detection and removal).

---

[13] TCTF is a field in the header of the MAC PDU that allows the identification of the logical channel class on FACH and RACH transport channels (i.e., TFTF specifies whether the PDU carries BCCH, CCCH, CTCH, SHCCH or dedicated logical channel information).

- *Add* the UE identifier for transmissions on RACH to identify the data from this UE;

- *Select* the TF (*Transport Format*), on the basis of the TFCS defined by the RRC. The scheduling/management of the logical channel priorities is related to the TF.

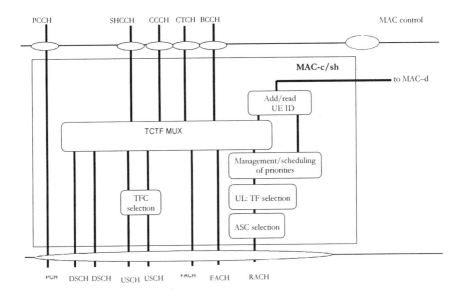

Fig. 18: MAC-c/sh architecture (UE side). ASC (*Access Service Class*) is used for the random access scheme on PRACH and will be described in the following sub-Section 5.3.1.

The RLC provides RLC-PDUs to the MAC, which fit into the available transport blocks on transport channels.

MAC-d entity - UE side

The MAC-d entity is responsible for mapping dedicated logical channels for uplink either into dedicated transport channels or to transfer data to the MAC-c/sh to be transmitted through common channels (a dedicated logical channel can be simultaneously mapped onto DCH and DSCH).

MAC-c/sh entity - UTRAN side (in every cell)

Fig. 19 shows the MAC-c/sh entity, UTRAN side, that has the following functionalities:

- *Scheduling/management of priorities*: management of FACH and DSCH resources among data flows, according to their priorities;

- Check the traffic volume through the *TCTF field*;

- *TFC selection* in downlink for FACH, PCH and DSCHs;

- *De-multiplexing* to separate USCH data from different UEs to be transferred to the different MAC-d entities;

- *Downlink code allocation*, a function that indicates the code used on DSCH.

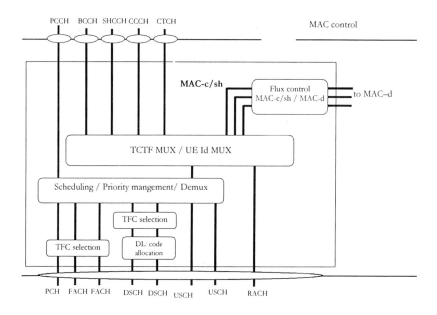

Fig. 19: MAC-c/sh architecture (UTRAN side).

MAC-d Entity - UTRAN side

The MAC-d entity on the UTRAN side offers the following additional functionalities with respect to the MAC-d entity on the UE side:

priority setting on the data received from dedicated control and traffic logical channels (DCCH/DTCH); downlink scheduling/priority management on the basis of the allowed TFCS assigned by RRC; flow control.

5.3.1   MAC services and functions

The MAC sub-layer provides the following services to higher layers:

- *Data transfer*. This service provides unacknowledged transfer to MAC PDUs between peer MAC entities.

- *Reallocation of radio resources and MAC parameters*. This service executes, upon request of RRC, the reallocation of radio resources and the modification of MAC parameters (reconfiguration of MAC functions such as change of UE identity, modification of transport format sets and change of transport channel type). The MAC can autonomously carry out resources reallocation.

- *Reporting of measurements*. Local measurements such as traffic volume and quality indication are reported to RRC.

An important MAC function is the traffic scheduling according to priorities. Another function is the *Access Service Class* (ASC) selection for transmissions on the Physical RACH (PRACH) resource, typically on one slot of the frame. An ASC entails a group of channelization codes, PRACH sub-channels and a permission to transmit [76]. In fact, each PRACH slot is divided among sub-channels; each sub-channel has occurrence every 1, 2, 4 or 8 frames. Once the random access attempt is permitted by the transmission probability, a sub-channel and a channelization code are randomly selected among those assigned to the ASC in order to send an access request.

The MAC layer maps logical channels on transport channels (see Fig. 17). Logical channels can be classified into two groups:

- *Control channels* for the transfer of control-plane information;

- *Traffic channels* for the transfer of user-plane information.

5.4   RLC services and functions

The RLC sub-layer provides the following services to higher layers:

- *Transparent data transfer*: transmission of SDUs of higher layer protocols without adding any protocol information.

- *Unacknowledged data transfer*: transmission of SDUs of higher layer protocols without guaranteeing the delivery to the peer entity. This service has the following characteristics:
  - *Detection of erroneous data*: the RLC sub-layer shall deliver SDUs to the receiving higher layer protocol only if they are free of transmission errors verified by using the sequence number check function.
  - *Immediate delivery*: the receiving RLC sub-layer entity shall deliver an SDU to the receiving higher layer protocol as soon as it arrives at the receiver.

- *Acknowledged data transfer*: the transmitter is informed if the RLC block is not correctly received. Such transfer mode has the following characteristics:
  - *Error-free delivery* by means of retransmissions.
  - *Univocal delivery*: the RLC sub-layer shall deliver each SDU only once to the receiving higher layer protocol by using the duplication detection function.
  - *In sequence delivery*: the RLC sub-layer provides for the delivery of SDUs in the right sequence.
  - *Out-sequence delivery*: the receiving RLC entity may deliver SDUs to higher layers protocols in a different sequence with respect to that submitted to RLC at the transmitting side.

- *Maintenance of QoS as defined by higher layers*: the retransmission protocol shall be configurable from layer 3 to provide different QoS levels.

- *Notification of unrecoverable errors*: RLC reports to the higher layer protocols the occurrence of errors that cannot be corrected by RLC through the normal procedures of exceptions management (e.g., modification of the maximum number of retransmissions according to delay requirements).

Note that there is a single RLC connection per radio bearer.

RLC functions are as follows:

- *Segmentation and reassembly.* This function performs segmentation/reassembly of variable-length higher-level SDUs into/from smaller RLC PDUs. The RLC PDU size is adaptable to the real set of transport formats.

- *Concatenation.* In the transmission of RLC PDUs, the first segment of the next SDU can be concatenated with the last segment of the previous SDU.

- *Padding.* When concatenation is not applicable and the remaining data to be transmitted does not fill an entire RLC PDU of a given size, the data field shall be filled with padding bits.

- *Transfer of user data.* It is used for the transfer of data between users of RLC services. RLC supports acknowledged, unacknowledged and transparent data transfer.

- *Error correction.* This function provides error correction by means of retransmissions (e.g., *Selective Repeat, Go Back N, Stop-and-Wait* ARQ) for the acknowledged transfer mode.

- *In sequence delivery of higher layer SDUs.* This function preserves the order of higher layer SDUs sent by RLC with the acknowledged transfer service. If this function is not used, out-sequence delivery is adopted.

- *Duplicate detection.* This function detects duplicated received RLC PDUs and assures that the resulting higher layer SDU is delivered only once.

- *Flow Control.* This function allows an RLC receiver to control the rate at which the peer RLC entity transmitting entity can send information.

- *Sequence number check.* This function is used in acknowledged mode and guarantees the integrity of reassembled SDUs and provides a mechanism for the detection of corrupted RLC PDUs through checking the sequence numbers of RLC PDUs when they are reassembled into SDUs.

- *Protocol error detection and recovery.* This function detects and recovers from errors in the operation of the RLC protocol.

- *Ciphering.* This function avoids unauthorized acquisition of data. The ciphering is executed in the RLC layer for non-transparent RLC mode.

- *SDU discard.* This function allows an RLC transmitter to discard RLC SDUs from the buffer.

The following description is focused on packet-switched downlink transmissions, a critical part of the air interface, especially in the presence of multimedia (video streaming and Web bursty) traffics that typically are asymmetric. In particular, we address the use of the DSCH transport channel (the corresponding uplink channel for packet-data traffic is the USCH; note that also RACH and FACH can be used for sporadic data traffics).

5.5   Resource management for DSCH

DSCH is a downlink transport channel (see Fig. 15) mapped to one or more physical channels. The two following cases are supported by UMTS Release '99, respectively referred to as case A and case B [72]:

- ***Case A***: DSCH is established as an extension to DCH transmissions. DSCH resource assignation is signaled by using the field for the transport format allocation (TFI), mapped to the TFCI of the associated DCH.

- *Case B*: DSCH is defined as a shared downlink channel for which the resource allocation is performed at the RRC layer in the *Controlling RNC*. The allocation messages, including UE identification, are transmitted on SHCCH, mapped on RACH/FACH. Several DSCH can be multiplexed on a CCTrCH in the physical layer. The TFs of the DSCHs have to be selected according to the TFCS of this CCTrCH. Every CCTrCH is mapped on one or more PDSCHs (i.e., the physical resource units corresponding to DSCH). If the TFCS of a CCTrCH contains more than one TFC, a TFCI can be transmitted inside the PDSCH or "blind" detection can be adopted by the UE.

Cases A and B can be simultaneously employed on an individual PDSCH.

Interleaving for the DSCH can be applied on several frames. However, the basic case is rectangular interleaving for a given PDU and corresponding to one radio frame (10 ms). A PDSCH can be assigned to different UEs at each new frame (MAC multiplexing).

In case A, transport blocks on the DSCH can be of fixed size, so that the TBS can be derived by the code assigned to each UE on the DSCH. In case B, TFCS can change every time resources are reallocated (from 1 frame to a maximum time of 8 frames).

5.5.1   Resource allocation and UE identification on DSCH

Referring to downlink transmissions on DSCH, the following resource allocation schemes can be considered.

Case A (UE requires a downlink TFCI on a *Dedicated Physical Control CHannel,* DPCCH): The TFCI of the dedicated physical channel can convey the information that a given code of the DSCH must be intercepted by the UE. Fast power control can be applied per code, based on the dedicated physical control channel (DPCCH). Alternatively, a UE may be requested on the DCH to listen to a DSCH for a time period to decode its address.

Case B (UE requires a downlink SHCCH): The information on *which* physical downlink shared channel has to be listened and *when* is sent by RRC on the SHCCH logical channel (mapped on RACH, USCH/FACH and DSCH). The transmitted Layer 3 messages contain information about the used PDSCHs and the timing of the allocation.

5.5.2   DSCH model in UTRAN

As it can be shown in Fig. 20, two RLC parts (i.e., two UE downlink traffic flows) point to the same DSCH, that is a shared resource; both cases A and B are addressed [72].

As shown in Fig. 20, the MAC sub-layer of a DSCH (UTRAN side) is divided between *Controlling RNC* (C-RNC) and *Serving RNC* (S-RNC). S-RNC supports the slow *Dynamic Channel Allocation* (slow DCA) and surveys the resource management related to more Node Bs (e.g., during a soft handover procedure); moreover, S-RNC also supports call admission and handoffs (if another RNC is involved in the active transmission through an inter-RNC handoff, it is named *Drift RNC,* D-RNC). Whereas, C-RNC supports the fast dynamic traffic scheduling (fast DCA), by updating resource allocation on a frame basis (or at most every 8 frames). Note that Iur is the interface between RNCs. For a given UE, C-RNC and S-RNC can be separate RNCs.

Each UE has terminated the RLC sub-layer in its S-RNC. Since Iur can support more DSCH data streams, the UEs on the DSCH can be related to different S-RNCs. The MAC layer in the network is responsible for mapping downlink data to a common channel (FACH, not shown in Fig. 20) or to a DCH and/or to the DSCH.

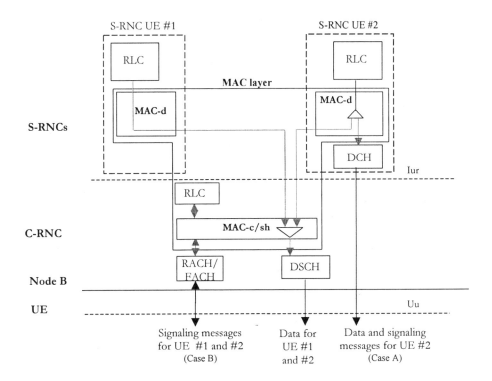

Fig. 20: Model of the downlink shared channel (DSCH) in UTRAN.

An analogous description to that shown in Fig. 20 can be used for USCH transmissions.

5.6   Performance evaluation for packet traffic over UTRA-TDD

The frame T_f of 10 ms is divided into 15 time slots T_s each of 667 µs. Each slot can be allocated to either uplink or downlink [67],[136]. Different slot assignments to uplink and downlink are possible (traffic asymmetry is supported). Within each time slot different transmissions are distinguished by means of orthogonal channelization codes and Node B signature codes (downlink) or UEs signature codes (uplink).

We refer here to a *Resource Unit* (RU) as the smallest allocable physical resource unit that consists of a spreading code on a given time slot. Spreading codes belong to the OVSF family. We envisage only packet data traffics (i.e., packet-switching).

In this study we make the following preliminary assumptions (see Fig. 21): (*i*) there is a single switching point between uplink and downlink in the frame (downlink slots are followed by uplink ones); (*ii*) the first downlink slot is destined to broadcast common and signaling channels (i.e., the beacon signal of the cell); (*iii*) the first uplink slot is used by the *Physical Random Access Channel* (PRACH); (*iv*) two traffic classes are considered, that is video conversational traffic (class #1) and Web interactive traffic (class #2).

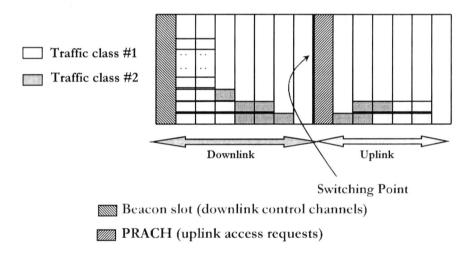

Fig. 21: Assumed frame organization for UTRA-TDD.

We have adopted a simplified access scheme with respect to that described in the previous Section 5.3.1. In particular, we consider:

- Two types of access bursts for PRACH: one with the guard period at the beginning of the burst and the other with guard period at the end of the burst.

- UEs needing to send requests to the Node B select at random both the code (16 codes/PRACH slot) on the PRACH slot and the burst type.

- A collision occurs when two UEs decide to transmit on the PRACH by using both the same code and the same burst type. Colliding UEs on the PRACH of a frame make a new attempt on the PRACH of the next frame.

We refer here to the UTRA-TDD burst type 2 format with two data fields of 1104 chips, midamble of 256 chips and guard period of 96 chips.

We recall that downlink transmissions only use the maximum PG value of 16 to facilitate the implementation of low-cost terminals. More downlink transmissions to the same UE are allowed by using different codes per slot (i.e., different RUs/slot). Each UE can use at most two different simultaneous channelization codes for uplink transmissions (i.e., two codes per slot or, equivalently, 2 RUs/slot) [69]. Concentrating the transmissions of a UE on a slot with different OVSF codes permits to reduce the intra-cell interference experienced by this UE. In general, the number of RUs that a UE can use (uplink/downlink) in each slot depends on many factors, such as number of OVSF codes, required QoS, interference levels and allowed power levels.

We consider that the Node B (or the C-RNC) is able to perform at the MAC level a fast scheduling of traffics on the available RUs so as to fulfill power and QoS constraints (see Fig. 22). Hence, a fast dynamic resource allocation is performed on a frame basis for both uplink and downlink (slow dynamic channel allocation is performed at the RNC level in order to make long-term scheduling decisions, such as: number of slots per frame assigned to circuit-switched traffics and number of slots assigned to packet-switched traffics; subdivision of resources between uplink and downlink) [136].

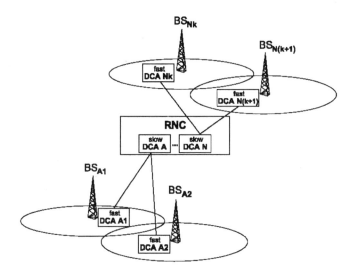

Fig. 22: Dynamic channel allocation with UTRA-TDD.

5.6.1   Study assumptions

We refer to a microcellular scenario with *Video Terminals* (VTs) and *Web browsing Terminals* (WTs).

In uplink, all the video (Web) traffic sources are ideally power controlled so that each of their packets is received with power level P_v (P_w). In downlink, another power control scheme applies to define the power levels of the transmitted by the Node B to different UEs, according to the power required by the most attenuated user signal.

Referring to uplink, we describe below the RU allocation constraints for what concerns QoS, available power and OVSF code orthogonality condition. Similar conditions can be derived for downlink transmissions.

QoS allocation constraints

It has been shown that the transmissions in surroundings cells increase the average interference level at the Node B. Assuming a constant number M of users in every cell, the inter-cell interference produced by transmissions in other cells (thus, power controlled by other Node Bs), can be modeled as the contribution of $M\varepsilon$ equivalent users in the same cell, where parameter ε has typically a value around 0.6 [12]. This consideration can be applied to the case of different traffic sources per cell, provided that all the cells have uniform traffic conditions.

With single-user receivers, the QoS in terms of BER is guaranteed for video and Web traffic transmissions on a slot if the following conditions are fulfilled for the received energy per bit-to-total noise spectral densities:

$$\left.\frac{E_b}{I_0}\right|_v \approx \frac{PG_v(1+r)P_v}{\left[M_v P_v + M_w P_w\right](1+\varepsilon) + \eta_0 W} \geq \left.\frac{E_b}{I_0}\right|_{v,req}. \tag{40}$$

$$\left.\frac{E_b}{I_0}\right|_w \approx \frac{PG_w(1+r)P_w}{\left[M_w P_w + M_v P_v\right](1+\varepsilon) + \eta_0 W} \geq \left.\frac{E_b}{I_0}\right|_{w,req}. \tag{41}$$

where:

M_v = Number of voice packets that can be allocated on a slot (i.e., RUs/slot)

M_w = Number of Web traffic packets that can be allocated on a slot (i.e., RUs/slot)

η_0 = Background noise power spectral density

W = Total spreading bandwidth (i.e., 3.84 *1.22 MHz, for a pulse shape roll-off r factor equal to 0.22)

P_v = Received power for a packet from a video source

P_w = Received power for a packet from a Web source

PG_v = Processing gain value for a video source

PG_w = Processing gain value for a Web source.

UEs update their transmission power levels on a frame basis according to the signal strength received from the beacon slot (open loop power control is effective due to the reciprocal nature of the TDD air interface).

On the basis of the above highlighted constraints, we consider our numerical assumptions. VT traffic is generally heavier than the WT one; therefore, we adopt: $PG_v \leq PG_w$. Moreover, since Web traffics (and data traffics, in general) typically need a greater protection than video ones, we consider here $E_b/I_o|_{w,req} \geq E_b/I_o|_{v,req}$, assuming for both VTs and WTs the same coding scheme with rate 1/3. Hence, we adopt $E_b/I_o|_{w,req} = 5$ dB and $E_b/I_o|_{v,req} = 3$ dB (conservative values that include a fading margin to reduce the risk of outage events).

OVSF orthogonality code constraints

Among the PG_w ($\geq PG_v$) spreading codes for Web traffic packets two codes are lost for OVSF orthogonality reasons for each packet transmitted with spreading $PG_w/2$. Hence, considering video packets with spreading $PG_v = PG_w/2^j$, with $j = 0, 1, ..., 3$, the OVSF orthogonality constraint is:

$$PG_w = M_w + \frac{PG_w}{PG_v} M_v. \qquad (42)$$

Moreover, we have the following code constraints from the 3GPP standard: in uplink transmissions, a VT or a WT cannot use more than two codes per slot; in downlink transmissions, $PG_v = PG_w = 16$.

Power allocation constraints

We assume in this study the *Outdoor-to-Indoor and Pedestrian test environment* shown in Chapter 2 of Part II [53]. Accordingly, path loss is characterized as in (23). For a microcellular scenario we will consider here a maximum cell radius R of 230 m. The maximum power, $P_{t,max}$, transmitted by a UE must be lower than 21 dBm with UTRA-TDD. Hence, the maximum received power, $P_{r,max}$, is:

$$P_{r,max} = P_{t,max} + G_{tx} - L + G_{rx} \text{ [dB]}. \qquad (43)$$

where G_{tx} (G_{rx}) is the antenna gain at the transmitter (receiver). We will assume for both uplink and downlink that $G_{tx} + G_{rx} = 11$ dB [137].

Note that if the UE transmits on two codes of the same slot, the maximum transmit power has to be divided between the two transmissions.

The received maximum power $P_{r,max}$ is a function of the distance. In this study we assume that all the UEs are at the same distance R from the Node B in order to evaluate the system performance in the worst-case. For $R = 230$ m, we have $L = 122$ dB and $P_{r,max} = -90$ dBm. Note that the received power level is much greater than the background noise power level $\eta_0 W$ (on the order of about -100 dBm). Hence, $\eta_0 W$ can be neglected in (40) and (41). It is easy to show that in such circumstances, the assumption of equal-distance R UEs has no practical impact in the constraints on M_v and M_w corresponding to (40) and (41), since they only depend on the ratio of power levels P_v and P_w, but not on their absolute values.

The resource allocation space

We name *admissible region* an area in the M_v-M_w plane where all the combinations of M_v and M_w values fulfill QoS, OVSF code orthogonality and maximum transmission power constraints. Fig. 23 has highlighted the regions where the QoS conditions (40) and (41) are met for different choices of the ratio between P_v and P_w and with $PG_v = PG_w = 16$. Moreover, it is easy to show that the constraint on the maximum available transmission power has no impact on the above regions for video and Web traffic packets. Finally, the QoS constraint is more stringent than the OVSF code orthogonality since the condition $M_v + M_w \leq 16$ is fulfilled in the highlighted regions. On the basis of the results in Fig. 23, we may conclude that the best solution is to select $P_w/P_v = 2$, since, in this case, a larger admissible region can be achieved.

Quite larger admissible regions with respect to those shown in Fig. 23 could be obtained by relaxing the target values $E_b/I_o|_{w,req}$ and $E_b/I_o|_{v,req}$. Hence, the study made here can be considered as a worst-case analysis, where QoS requirements strongly limit the system capacity.

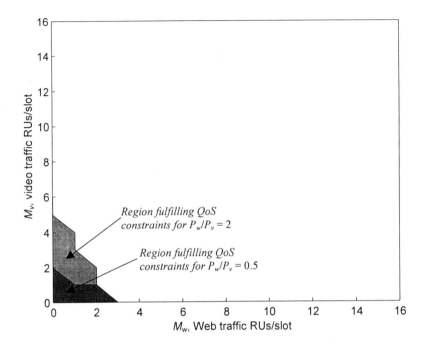

Fig. 23: Regions fulfilling the QoS constraints and admissible regions for $PG_v = PG_w = 16$ (simplified case with roll-off factor equal to 0).

5.6.2   The proposed RRM scheme

We describe here our proposed packet scheduling algorithm operated at the Node B (or C-RNC) level that assigns slot-code RUs to uplink (and downlink) traffics of the cell on a frame basis. This scheme has been generically named, *Fast-Dynamic Resource Allocation* (F-DRA) scheme.

Two traffic classes have been envisaged: video and Web traffic sources. We describe here the uplink F-DRA scheme based on the allocation constraints previously outlined, but similar considerations can be drawn for the management of downlink traffics.

Uplink traffic sources have to inform the Node B of their transmission needs by means of explicit requests. UEs that have not an open session with the Node B adopt a random access phase on the PRACH as soon as they need to transmit data. The access phase with contentions is shared by all traffic classes. We assume that UEs with already opened sessions adopt a piggybacking scheme (on an associated control channel) to update their transmission requests at the Node B. Each request (through either an access burst or the piggybacking scheme) is supposed to include: terminal ID, traffic class and number of packets to be transmitted.

On the basis of the requests coming from the UEs, the F-DRA scheduler decides the assignment of the different codes of the slots in the uplink part of the frame by fulfilling on each slot the allocation constraints described in the previous sub-Section 5.6.1. Scheduling decisions taken during the uplink period will be broadcast to UEs in the first downlink beacon slot of the next frame.

The F-DRA scheduler implements the following criteria:

1. It is better to assign the traffic of a given UE in the lowest possible number of slots to reduce the experienced intra-cell interference level. Moreover, the traffic of a user experiences lower delays if we try to compact it in few slots.

2. The requests coming from different MTs are served according to an RR scheme.

3. Real-time video traffic requests are always served before Web traffic requests that need high reliability, but can tolerate some transmission delays.

5.6.3 Simulation results

On the basis of the previous assumptions, simulations have been carried with the models shown in Chapter 2 in Part II to evaluate the performance of the F-DRA scheme.

As for VTs we use the D-MAP videoconference traffic model with $\mu = 144$ kbit/s, $\sigma = 64$ kbit/s, $M = 5$ minisources/VT and $a = 3.9$ s^{-1} [114]. On the basis of these values, the minisource activity factor ψ_v is equal to 0.5. As for WT interactive traffics we have adopted the Web traffic model in Chapter 2 of Part II with $q = 8$, for a high burstiness traffic degree [53]. As shown in Fig. 21, we consider that the first 8 slots are used for downlink, whereas the next 7 ones are destined to uplink A coding scheme with rate 1/3, a burst type 2 and processing gains $PG_v = PG_w = 16$ have been selected. Consequently, the information content conveyed by a burst (i.e., RU transmission content) has been determined: the bits generated by VTs or WTs sources are packetized according to the burst payload.

The video packet dropping probability $P_{drop,v}$ cannot exceed a maximum value here relaxed to 10^{-2} with $D_{vmax} = 90$ ms. The QoS parameter here evaluated for WTs is the *mean packet transmission delay*, $T_{d,pkt}$, (i.e., the mean delay from the packet generation to the instant when its transmission is completed).

Each simulation result shown below has been obtained by repeating 10 times a simulation of 50 millions of slots in order to obtain reliable results. We consider two tiers of cells surrounding a central cell; each cell has the same traffic conditions and radius 230 meters. Figs. 24 and 25 respectively show $P_{drop,v}$ and throughput (in pkts/RU) results for uplink as a function of the number of video and Web traffic sources. We can note that a $P_{drop,v}$ value lower than 10^{-2} can be guaranteed only if the system load is low (i.e., at most 2 uplink video sources). In these circumstances, the total throughput achieved by F-DRA varies between 0.3 and 0.5 pkts/RU depending on the number of Web traffic sources (see Fig. 25). The low capacity values obtained here are related to the conservative assumptions we have made in terms of PG values and QoS requirements.

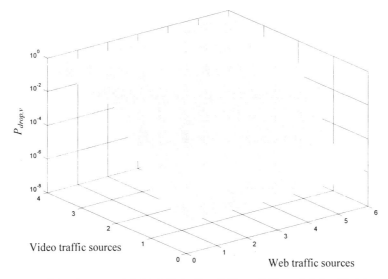

Fig. 24: $P_{drop,v}$ with F-DRA for $PG_v = PG_w = 16$ (uplink).

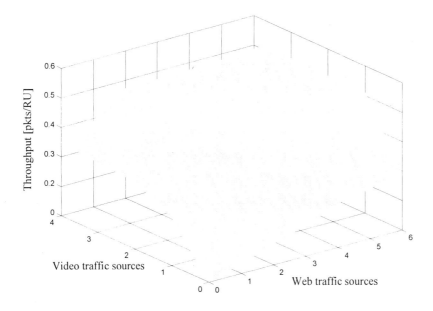

Fig. 25: Throughput with F-DRA for $PG_v = PG_w = 16$ (uplink).

The same RRM scheme can be applied to both uplink and downlink. In downlink we have significant advantages that allow reducing delays due to the following aspects: (*i*) there is not the contention phase to notify transmission needs; this causes lower delays; (*ii*) downlink traffics experience negligible intra-cell interference due to the fact that they are synchronized (transmissions from the same Node B); (*iii*) we have assigned one more slot to downlink with respect to uplink.

The results of the average packet delay of both video and Web traffics versus the throughput in pkts/frame are shown in Fig. 26 for both downlink and uplink. In performing these simulations we have increased the number of video sources in the presence of a fixed number of Web sources (for both uplink and downlink). As expected, delays increase with throughput. Moreover, downlink traffics experience lower delays than uplink ones. Finally, data packet delays are greater than video ones, since Web traffics have a lower priority.

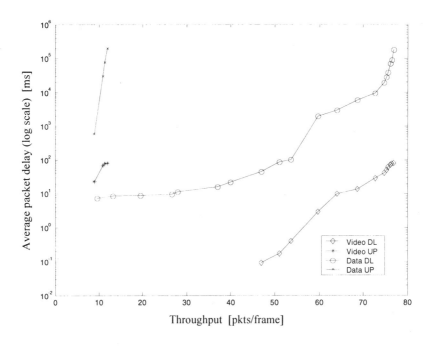

Fig. 26: Average packet transmission delay with F-DRA for $PG_v = PG_w$ = 16.

Chapter 6: RRM in wireless microcellular systems

We refer here to a *wireless mobile Asynchronous Transfer Mode* (wmATM) scenario, where users with MTs run real-time applications (e.g., video) and Internet applications (e.g., Web surfing). We consider a broadband wireless system operating in the 5.2 GHz, where MTs are managed by *Access Points* (APs) providing pico- and micro-cellular coverage. A layer 2 equivalent channel bit-rate B_c = 10 Mbit/s has been envisaged in both uplink and downlink according to a TDMA-FDD air interface. Each packet (*extended ATM cell*) is supposed to contain 54 bytes with payload of L_p = 48 bytes. The packet transmission time is $T_{pkt} = 8L/B_c$.

We propose here a novel RRM scheme where MTs are cyclically authorized to transmit by the AP according to a *polling scheme* (see Fig. 27). Correspondingly, an RR scheme should be used for downlink transmissions.

Within the population MTs of a cell we have M_v *Videoconferencing Terminals* (VTs) and M_w *Web browsing Terminals* (WTs).

Our interest for polling/RR techniques has been motivated by the IEEE 802.12 *AnyLan* standard for wireline networks that envisages two levels of priorities for multimedia traffics. We consider a similar approach for a wireless scenario, where VTs are prioritized with respect to WTs. The polling/RR approach is particularly attracting for managing heavy traffic sources producing real-time traffics, since it guarantees a maximum packet transmission delay and, therefore, permits to control the probability that a packet is not served in time. Polling/RR schemes are also fair resource allocation techniques, since they allow a minimum bit-rate for all the sources. Anyway, a good fairness level can be achieved by integrating, as proposed here, polling/RR with a token bucket policer to regulate adaptively the maximum number of packets that a source is enabled to transmit at each cycle.

A priority order is adopted in the polling cycle in order to privilege both real-time traffic sources and the VTs with the most congested buffers.

For all these reasons, our proposed RRM scheme has been named *Adaptive Token Bucket, priority-based Polling* (ATB-P). This technique has been applied to uplink transmissions, but the same scheme could be directly adopted for downlink, according to the FDD air interface assumption.

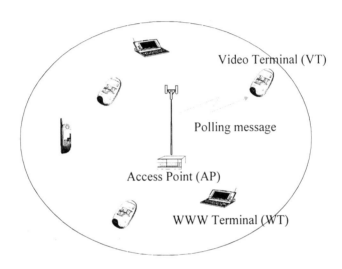

Fig. 27: The envisaged wireless scenario with the AP that enables MTs to transmit according to a polling scheme.

Our polling-based MAC protocol can be considered as an improvement of the scheme proposed in [138], where the AP enables the transmission of one packet per cycle for each source. Such strategy may lead to high polling overheads (low protocol efficiency) in the presence of bursty traffic sources, as those considered here. Otherwise, the ATB-P scheme uses a scheduler at the AP that decides both the source service order and the transmission of more than one packet per source at each cycle on the basis of minimum information received from MTs.

Another interesting polling technique has been shown in [139], where an adaptive technique is envisaged: MTs are polled depending on their bit-rate and the time elapsed since their last poll. Our ATB-P technique further develops this idea by integrating the polling scheme with a token bucket regulator that estimates the resources to be assigned to an MT in each cycle.

In our scenario, the polling scheme is particularly efficient since:

- Round-trip propagation delays are negligible with respect to the packet transmission time (in our examples, $T_{pkt} \approx 4.32 \times 10^{-5}$ s, whereas the maximum round-trip propagation delay in a micro-cell of radius 300 m is approximately equal to 2×10^{-6} s);

- Polling messages are designed to entail low overhead;

- VTs produce a varying but generally heavy traffic: due to the cycle priority order, the probability that the VT has an empty buffer when it is served is negligible.

We measure the packet dropping probability due to deadline expiration, $P_{drop,v}$, for VTs and the mean packet transmission delay $T_{d,pkt}$, for WTs.

6.1   ATB-P protocol description

For each source, the AP adopts a token bucket regulator with depth B where tokens are feed with a rate of r tokens/s (see Fig. 28).

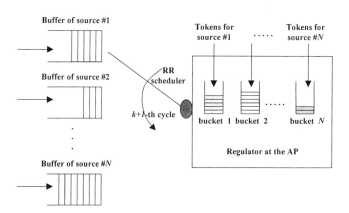

Fig. 28: Token bucket regulator and RR scheduler at the AP to manage uplink transmissions of different traffic sources.

The AP cyclically triggers the transmissions of an MT by sending a polling message that contains n tokens (= bucket content for the MT), thus enabling the MT to send up to n packets to the AP. If the MT has no packet to transmit, it immediately sends a release command so that the AP enables the next MT in its list. The last packet transmitted by an MT piggybacks (in case of token exhaustion) in an appropriate header field the residual number of packets still in the MT buffer.

At the end of the k-th cycle the AP evaluates the parameters in Table 4 for each source.

Symbol	Definition
$t^k_{arr,i}$	Time elapsed from the beginning of the k-th cycle and starting time for the service of the i-th MT in the k-th cycle.
t^k_i	The service time for the i-th MT in the k-th cycle
τ^k_i	Time elapsed from the service end of the i-th MT in the k-th cycle to the beginning of the new $(k+1)$-th cycle
δ^k_i	The interrogation time for the i-th MT in the k-th cycle (i.e., all overhead related to the i-th MT in the k-th cycle, including interrogation time and messages sent by the MT to notify its buffer status)
G^k_i	Number of packets arrived at the i-th MT in the interval from the beginning of its service in the $(k-1)$-th cycle to the beginning of service in the k-th cycle.
R^k_i	Residual number of packets left in the buffer of the i-th MT at the end of the service in the k-th cycle due to token exhaustion.

Table 4: Description of the polling variables.

Note that G^k_i and R^k_i values are notified by the MT itself at the beginning and at the end of its service in the k-th cycle (piggybacking scheme). Index i corresponds to VTs for $1 \le i \le M_v$ and to WTs for $M_v + 1 \le i \le M_v + M_w$. According to Table 4, the duration of the n-th cycle, σ^n, and the time elapsed up to the k-th cycle, Σ^k, are:

$$\sigma^n = \sum_{i=1}^{M_v+M_w}\left(t_i^n + \delta_i^n\right) \text{ [s]} \quad \text{and} \quad \Sigma^k = \sum_{n=1}^{k}\sigma^n \text{ [s]} . \tag{44}$$

The AP estimates the token rate T_i^k for the generic i-th MT to be used in the $(k+1)$-th cycle, as:

$$T_i^k = \begin{cases} \dfrac{\sum\limits_{n\in\Omega_w^k}G_i^n}{\Gamma_\Delta^k}, & \text{for VTs} \\[2em] \dfrac{\sum\limits_{n=1}^{k}G_i^n}{\Sigma^k}, & \text{for WTs} \end{cases} \qquad \left[\dfrac{\text{cells}}{\text{s}}\right] \tag{45}$$

where Γ_Δ^k is the time on which the packet generation rate is evaluated for VTs and depends on both the cycle index k and Δ, the *coherence time* for the VT (see (47)), as follows:

$$\Gamma_\Delta^k = t_i^h + \tau_i^h + \sum_{n=h+1}^{k-1}\sigma^n + t_{arr,i}^k :$$

$$t_i^h + \tau_i^h + \sum_{n=h+1}^{k-1}\sigma^n + t_{arr,i}^k \le \Delta < t_i^{h-1} + \tau_i^{h-1} + \sum_{n=h}^{k-1}\sigma^n + t_{arr,i}^k , \tag{46}$$

$$\text{for } h \le k .$$

Moreover, the set Ω_W^k is equal to $\{h, h+1, ..., k\}$.

Let Δ denote the mean sojourn time in a state, evaluated here referring to the D-MAP video traffic model described in Chapter 2 of Part II:

$$\Delta = \sum_{j=0}^{M}\frac{1}{j\alpha + (M-j)\beta}\,Prob.\{state = j\} \text{ [s]} . \tag{47}$$

According to the values assumed for VTs in Chapter 2 of Part II, we have: $\Delta = 65$ ms. At the beginning of the $(k+1)$-th cycle, the AP computes the priority order P_i^{k+1} to be adopted to serve VTs in the $(k+1)$-th cycle as:

$$P_i^{k+1} = R_i^k + T_i^k \tau_i^k, \quad 1 \le i \le M_v. \tag{48}$$

Parameter P_i^{k+1} represents an estimate of the number of packets in the buffer of the i-th VT at the beginning of the $k+1$-th cycle; such estimate is reasonable since the D-MAP video modulating process correlates the video traffic. In the $k+1$-th cycle, VTs are served in order of decreasing P_i^{k+1} value: the AP adapts the cycle order to serve first the VT with a greater buffer congestion. This heuristic *buffer-based* priority order is related to the assumption that the most congested buffer may experience packet dropping if not served in time [140].

Let B_v denote the bucket depth for VTs. After having served all VTs, the controller enables WT transmissions as follows: WTs can be served in the $k+1$-th cycle so that the *maximum cycle duration* is $B_v M_v$ packets. Hence, the maximum time available in the $k+1$-th cycle for WTs is:

$$T_{w\_max}^{k+1} = \begin{cases} B_v M_v T_{pkt} - \sum_{i=1}^{M_v} \{t_i^{k+1} + \delta_i^{k+1}\}, & \text{for } M_v > 0 \\ \text{no constraint}, & \text{otherwise} \end{cases}. \tag{49}$$

We have not considered a priority order among WTs. The AP enables the transmission of a WT by sending it the number of tokens, which is the minimum between the token bucket value and the remaining packets to reach the maximum cycle duration of $B_v M_v$ packets. If not all the WTs can be served in a cycle, the WT service restarts from the interruption point in the next cycle. The B_v value can be obtained by imposing that the maximum cycle duration cannot exceed the video packet deadline (hence, we can serve in time a packet generated by a VT soon after this VT has released the service):

$$B_v M_v T_{pkt} = D_{vmax}. \tag{50}$$

With the used parameter values, we have $B_v = 300$ packets for $M_v = 7$ VTs. Condition (50) entails that there is a maximum cycle duration that is independent of M_v.

6.2 ATB-P performance evaluation

VT traffic has been generated according to the fluid-flow video D-MAP model described in Chapter 2 of Part II [114] with the parameter values shown there. As for WTs, we have adopted the model proposed in Chapter 2 of Part II [53] with $q = 4$ and $m = 6666$ bytes. The ATB-P resource management scheme has been implemented in a simulator according to the values shown in Table 5. Very long simulations have assured reliable results.

Parameter	Value
Channel bit-rate	$B_c = 10$ Mbit/s
Packet format	Payload/total packet length = 48/54 bytes
Number of minisources per VT	$M = 10$ minisources/VT
Mean VT bit-rate	$\mu = 512$ kbit/s
Standard deviation of the bit-rate of a VT	$\sigma = 256$ kbit/s
VT auto-covariance parameter	$a = 3.9$ s^{-1}
Mean reading time of the WT	$m_{Dpc} = 4$ s
Mean number of datagrams per packet call	$m_{Nd} = 25$ datagrams/packet call
Mean datagram interarrival time in the packet call	$m_{Dd} = 0.125$ s (i.e., $q = 4$)
Maximum distance from AP	300 m
Polling overhead per MT	6 bytes

Table 5: System parameter values.

We compare the ATB-P scheme with the *Mobile Access Scheme based on Contention and Reservation for ATM* (MASCARA) with *Prioritized Regulated Allocation Delay Oriented Scheduling* (PRADOS) [105]. The MASCARA-PRADOS protocol is based on a variable-length frame formed by a downlink period, an uplink period and a contention interval. The AP defines the duration of these intervals at the beginning of each frame and dynamically schedules transmissions on the basis of the priority class, traffic characteristics and QoS requirements (e.g., maximum packet delay) declared by a connection in the set-up phase. The MT transmission needs are notified to the AP by means of requests (sent through the contention phase or piggybacked in the transmitted packets) that contain the number of packets in the MT buffer. The scheduler adopts the following service priority ([14]): CBR > rt-VBR > ABR > UBR. A leaky-bucket scheme is used for every uplink and downlink traffic source. Within a service class, traffic sources are served according to a round robin scheme. The protocol initially schedules the transmission of a packet just before its deadline; after having allocated all the packets, the scheduler anticipates the packet transmission time if some slots are idle.

Moreover, our scheduling approach will be validated through the comparison with the benchmark *Earliest Deadline First* (EDF) scheduling scheme, applied ideally as if the traffics of all the VTs were collected by the same transmission queue (*urgency-based* priority) [6], thus neglecting switching times from the service of an MT to another. EDF entails a dynamic priority scheduler that selects for transmission the packet with the shortest residual lifetime. The dynamic nature of the EDF scheme is evident from the fact that the priority of a packet increases with the amount of time it spends in the system. The EDF policy is optimal for minimizing the maximum packet transmission delay, thus providing the best $P_{drop,v}$ value "averaged" over all the VTs. In a practical case, the distributed nature of the traffic sources makes it impossible to reach a coordination like that envisaged by the ideal EDF case, for two reasons: (*i*) the scheduler cannot instantaneously know the present status of all the buffers of the real-time sources; (*ii*) the service cannot be frequently switched from a source to another, otherwise the overheads typical of the polling message would vanish any optimal scheduling.

Fig. 29 presents the video $P_{drop,v}$ comparison between ATB-P, MASCARA-PRADOS and the ideal EDF scheme (lower bound). These results clearly highlight that the ATB-P scheme outperforms the MASCARA-PRADOS approach due to improved ATB-P scheduling capabilities achieved by integrating a traffic estimator, a token bucket regulator and the polling scheme. The encouraging result is that below the video maximum $P_{drop,v}$ value of 10^{-4}, the difference between ATB-P and EDF reduces, thus validating our proposed scheme.

---

[14] According to the ATM terminology, we have: *Constant Bit Rate* (CBR), *real-time Variable Bit Rate* (rt-VBR), *Available Bit Rate* (ABR), *Unspecified Bit Rate* (UBR).

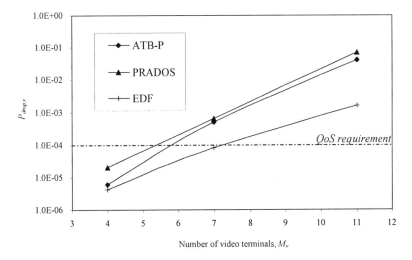

Fig. 29: Performance comparisons (video traffic only).

Figs. 30 and 31 respectively show $P_{drop,v}$ and $T_{d,pkt}$ as functions of the number of WTs for $M_v = 4$ VTs. We may note that the video traffic QoS is slightly sensitive to the presence of Web browsing traffic. Moreover, Web traffic experiences low delays if the number of WTs is sufficiently low.

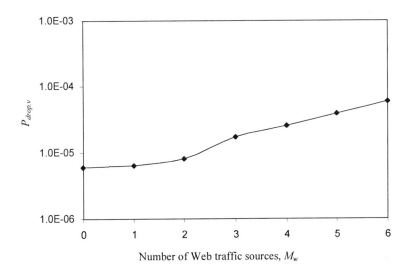

Fig. 30: $P_{drop,v}$ of VTs as a function of the number of WTs, M_w, for M_v = 4 VTs.

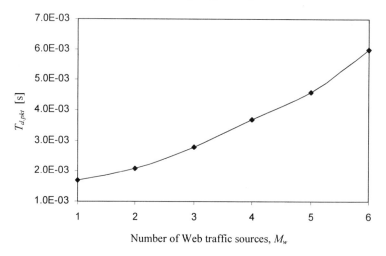

Fig. 31: $T_{d,pkt}$ of WTs as a function of the number of WTs, M_w, for M_v = 4 VTs.

Finally, the following Fig. 32 shows the delay performance for two different cases of data traffic sources with 10 VTs according to the model described in Chapter 2 of Part II [114]. In particular, we have considered the following data sources described in Chapter 2 of Part II: Web traffic sources (2-MMPP model with q = 4 [53]) or M/Pareto sources (r = 1.05 kbit/s, H = 0.9, λ = 3.5 bursts/s, δ = 0.5 s, γ = 1.2 [121]). Note that each M/Pareto source could model the aggregated traffic circulating on a *Local Area Network* that is collected and sent to an AP. 2-MMPP and M/Pareto sources have the same mean arrival rate of datagrams/bursts and datagrams have the same mean length of bursts. The delay measure taken in Fig. 32 refers to the datagram delay for the 2-MMPP source and to the delay to transmit all the packets of a burst in the M/Pareto case.

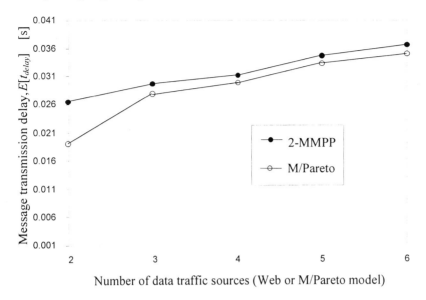

Fig. 32: Comparison of the ATB-P performance with 2-MMPP data
traffic sources or M/Pareto ones with 10 VTs.

We may note that the 2-MMPP model and the M/Pareto one have about
the same delay performance. Correspondingly, we have similar $P_{drop,v}$
values for both cases, around 10^{-3}.

Chapter 7: RRM in LEO-MSSs

The challenge of future mobile multimedia networks is to provide worldwide tetherless communication services. *Low Earth Orbit-Mobile Satellite Systems* (LEO-MSSs) will play a significant role by filling the coverage gaps of future-generation terrestrial cellular networks. We refer here to the management of uplink radio resources in the air interface between MTs and the satellite of a given constellation. Particularly efficient MAC protocols must be designed in order to share efficiently LEO satellite resources among users.

This Chapter deals with the definition of MAC schemes that are evolutions of techniques proposed for terrestrial micro-cellular systems, such as the classical *Packet Reservation Multiple Access* (PRMA) protocol [109] and the *Reservation Random Access* (RRA) scheme [110]. In both cases, a TDMA air interface is assumed. Let us focus on MAC integrating the management of isochronous and data bursty traffics.

All the MAC schemes described below assume that satellites have on board processing capabilities both to manage the transmission requests coming from MTs and to decide the TDMA resources to be destined to each of them.

7.1 The classical PRMA protocol in LEO-MSSs

PRMA is a decentralized-control protocol that combines random access (i.e., some form of *Slotted Aloha* scheme, where transmissions of available packets are controlled by permission probabilities) with slot reservation (i.e., a fixed use of resources, once their control has been acquired through a successful random access). The efficiency of PRMA relies on managing voice sources with *Speech Activity Detection* (SAD) as explained in Chapter 2 of Part II: only during a talkspurt, a voice source has reserved one slot per frame to transmit its packets. A feedback channel broadcast by the *cell controller* (i.e., the satellite in our case) informs the MTs about the state of each slot (i.e., *idle* or *reserved*) in a frame. As soon as a new talkspurt is revealed, the MT tries to transmit a packet in the first idle slot (*contending state*), according to a *permission probability* scheme. The satellite recognizes the request made by an MT by decoding the header of the received packet. When the transmission attempt of an MT is successful on a slot, the MT acquires the reservation of this slot in subsequent frames.

The main factor limiting PRMA performance in LEO-MSSs is the non-negligible *Round Trip propagation Delay* (RTD) value that prevents the MTs on the earth to know immediately the outcome of their transmission attempts. As shown in Part I of this book, RTD values vary from 5 ms to 30 ms in LEO-MSSs, depending on the satellite constellation altitude and the minimum elevation angle from MTs to the satellite. For a conservative study, RTD is here assumed always equal to its maximum value, RTD_{max}, for a given LEO-MSS.

In order to apply PRMA in a LEO-MSS, we consider that the frame duration T_f is equal to the $RTD_{max} + \varepsilon$, where ε is the packet header transmission time. Accordingly, after a successful attempt on a slot, the MT receives the outcome from the satellite before the homologous slot in the next frame. Hence, the MT knows in time, which is the slot to be used for transmissions. If more MTs attempt to send their packets on the same slot, there is a collision (unless capture phenomena occur): these MTs know that they must reschedule new transmissions only after RTD, as shown in Fig. 33. This delay reduces the efficiency of the PRMA scheme in LEO satellites, since voice packets not delivered within their deadlines are dropped, thus causing the annoying effect of *front-end clipping*.

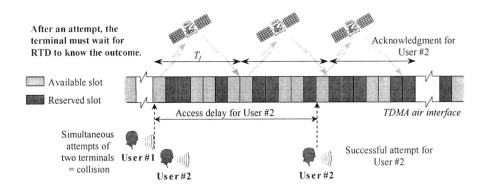

Fig. 33: Collision of two simultaneous attempts with PRMA in LEO-MSSs.

In order to relax the access delay problems due to RTD, several solutions can be considered that are also suitable for other scenarios where RTD is greater than the packet transmission time (e.g., HAPS). In particular, we consider below MAC schemes that derive from the modification of PRMA and one MAC scheme that derives from RRA.

In the following description, we will assume that M_s *Speech Terminals* (STs) and M_w *Web browsing Terminals* (WTs) need to be managed per carrier per cell.

7.2 *PRMA with Hindering States* (PRMA-HS)

In the classical PRMA scheme, assuming a typical LEO system, an ST could perform few access attempts before dropping the first packet of a talkspurt (the deadline of a real-time voice packet is typically equal to 32 ms). To overcome this problem, we consider a modification of the classical PRMA protocol: an MT is allowed to attempt transmissions (according to the permission probability scheme) also while it is waiting for receiving the outcome of a previous attempt [141]. Hence, if the previous attempt has been unsuccessful, this modification permits a faster access scheme. Whereas, if a previous attempt has been successful, further attempts are useless (they are discarded by the satellite) and may hinder the accesses to other MTs. This is a possible drawback of this scheme, but simulation results have shown that the advantages of a fast reattempt scheme overcome this problem due to useless contentions. Accordingly, this scheme has been called *PRMA with Hindering States* (PRMA-HS).

In order to integrate ST and WT traffics, different permission probabilities values have been considered; p_s and p_w, respectively, where $p_s > p_w$ to prioritize the voice real-time traffic.

7.3 *Modified PRMA* (MPRMA)

In the *Modified PRMA* (MPRMA) protocol, a given field in the packet header is devoted to notify the satellite if a transmission request (i.e., the first packet sent by an MT that must acquire transmission rights) comes from a WT or from an ST; consequently, different algorithms are used. In particular, the management of STs is as in the classical PRMA scheme. Whereas, WT requests are served by using a queue of transmission requests on board of the satellite.

If a message is generated by a WT when its buffer is idle, its first packet (= request packet) is transmitted at random on an available slot, according to a permission probability, p_w. Collisions may occur with the attempts of other MTs. Re-attempts are performed only after the RTD. As soon as the first packet of a message is correctly received (i.e., no collision occurs), its header contains a piggybacked WT transmission request to be stored in a service queue on the satellite. The controller on board of the satellite manages a queue of uplink transmission requests for each MPRMA carrier of a cell and decides allocations of slots (not used by STs) to WTs [142].

In order to guarantee a certain number of idle slots for new accesses of WTs and STs, the controller assigns an idle slot to a given WT (with its request at the head of the satellite queue) in the next frame, according to an access probability p_a.

An *exhaustive service policy* is adopted for the transmission of messages from a WT whose request is at the head of the service queue on the satellite. In fact, a suitable flag is set in the header of the WT transmitted packets to notify the request of further transmissions to the satellite (*piggybacked request*). Both the random access packet and the packet used for the piggybacked requests use an *ad hoc* field in the header to notify the message length to the satellite.

7.4   DRAMA protocol

We envisage a possible adaptation of the RRA scheme in the LEO-MSS scenario. Accordingly, we consider that the TDMA frame is divided in two separated parts (see Fig. 34): *access slots* and *information slots*. Access slots are minislotted: an MT sends an access burst on a minislot to request transmission resources. This new organization of the frame is adopted in order to avoid that wasted radio resources in the contention phase may affect the information throughput of the protocol. This technique has been named *Dynamic Resource Assignment Multiple Access* (DRAMA) [143]. The number of mini-slotted slots per frame is always greater than or equal to one (the number of minislots per slot is about equal to the number of packet headers that can be transmitted in a slot time). This value is dynamically updated trying to have a total number of minislots equal to the number of MTs that have set-up a connection (satellite-ATM) and that are not transmitting. This solution permits to maximize the throughput of successful requests per minislot. Colliding MTs will reattempt in the next contention phase. A feedback channel informs the MTs at the beginning of each frame about the number of minislotted slots for the next access phase.

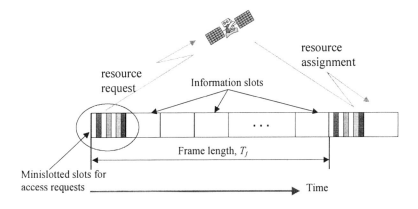

Fig. 34: The DRAMA organization of the TDMA frame.

With the DRAMA protocol the satellite has to manage distinct service queues for the different traffic classes (see the RRM approach proposed in Chapter 1 of Part II for the support of multimedia traffics). We consider here two queues: one for ST requests and the other for WT requests.

Slot allocations are dynamically updated. Slots are first assigned to fulfill ST requests. Remaining slots are used to serve WT requests. On the basis of currently active transmission requests, the satellite knows how many slots will be destined to STs in the next frame and decides slot assignments to data traffics by using a cyclic policy among all WTs. Each WT transmission request conveys the number of packets of the related message. Hence, the satellite may decide also multiple slot allocations to a WT in a frame (if room). The feedback channel is used to send assignment commands for each slot. Resource assignment commands are received by the MTs after a propagation delay.

The ST state diagram with the DRAMA protocol is shown in Fig. 34. State transitions may occur at the end of each slot. We may note that the ST is in the *Silent state* when no voice traffic is generated. As soon as a talkspurt is revealed, the ST waits for the next contention phase in the next frame. Hence, the ST enters the *Contending state*. In this state, the ST selects a minislot at random to transmit a minipacket with a request. If the transmission attempt is successful, the ST waits for the right transmission synchronism (and also for a free resource). As soon as a transmission request is received without collision, the ST is granted with the reservation of one slot per frame (if room). The WT state diagram is similar to that shown in Fig. 35, except for the fact that a queue state is considered on the satellite for the management of the transmission requests of the different messages.

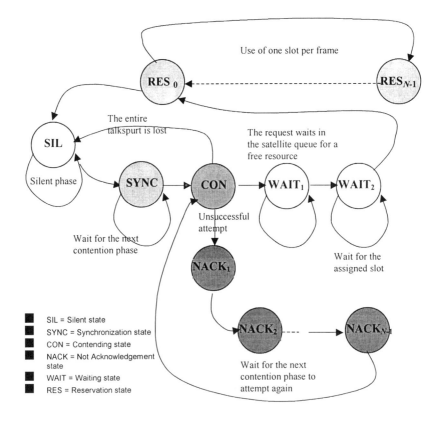

Fig. 35: ST state diagram (N denotes the number of slots per frame).

7.5   Performance comparisons

A simulator has been built to compare the performance of these MAC protocols in the presence of both ST and WT traffics. The typical ON-OFF voice source model described in Chapter 2 of Part II has been adopted for STs [109]. The 2-MMPP generator described in Chapter 2 of Part II has been employed for WTs [53]. We have evaluated the probability that a voice packet is dropped due to transmission deadline expiration, $P_{drop,s}$. We have also measured the mean transmission delay experienced by a datagram, $E[t_{delay}]$, for WTs.

The following simulated configurations are selected so that the offered total traffic load produced by all the STs and WTs is lower than 1 pkt/slot, so as to assure the *stability* of the transmission buffers of MTs.

Extensive simulation runs have been carried out to compare the proposed MAC schemes in a typical LEO satellite constellation at 780 km altitude with RTD_{max} = 15 ms. Simulation results have been obtained assuming: a channel bit-rate R_c = 765 kbit/s, a source bit-rate R_s = 32 kbit/s, a packet header H_p = 64 bits, a frame duration T_f = RTD_{max} = 15 ms. Moreover, for PRMA and PRMA-HS p_s = 0.6 and p_w = 0.2 [141]; for MPRMA p_s = 0.6, p_w = 0.4 and p_a = 0.8 [142]; for DRAMA, the number of minislots per slot is N_m = 8 [143]. Assuming a configuration with M_w = 12 WTs and M_s = 21 STs, simulation results have been shown in Figs. 36 and 37 for $P_{drop,s}$ and $E[t_{delay}]$, respectively. In abscissa we have used the total traffic load in Erlang due to WTs: $M_w\rho_w$, where ρ_w is defined in (7). The data traffic increase in abscissa has been obtained by progressively increasing q (i.e., q = 1, 2, 3, ...) that corresponds to a mean arrival rate of datagrams in the packet call state equal to $2q$ datagrams/s.

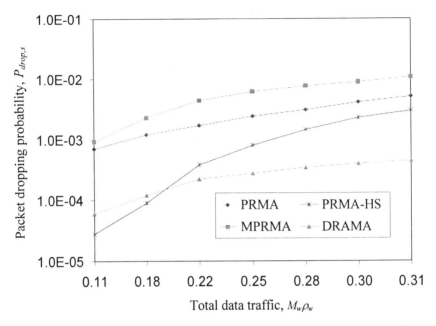

Fig. 36: Performance comparison among the considered MAC schemes in terms of $P_{drop,s}$.

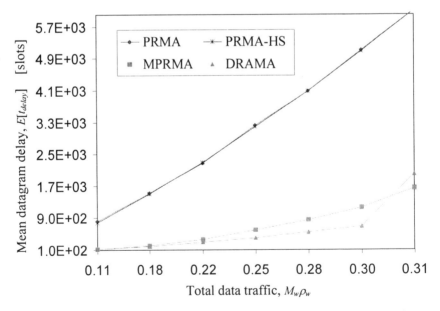

Fig. 37: Performance comparison among the considered MAC schemes in terms of $E[t_{delay}]$.

On the basis of these results, it is evident that MPRMA outperforms PRMA-HS in terms of $E[t_{delay}]$ and maintains $P_{drop,s}$ below the maximum acceptable value of 1%. However, the MPRMA scheme does not allow a satisfactory $P_{drop,s}$ performance. Hence, the novel DRAMA scheme has to be preferred, since it considerably reduces both $E[t_{delay}]$ and $P_{drop,s}$, thus achieving a better utilization of satellite resources with respect to PRMA, PRMA-HS and MPRMA.

The contention phase with the DRAMA protocol is quite efficient and does not reduce (significantly) the information throughput, especially in the most critical cases with high traffic loads. Hence, we have low access delays for new talkspurts, so that $P_{drop,s}$ is low.

Moreover, MPRMA and DRAMA techniques that may assign more than one slot per frame to a given WT (if available, after having managed ST traffics) allow a better service for WT bursty traffics, thus attaining lower $E[t_{delay}]$ values. This is a significant improvement with respect to PRMA and PRMA-HS that rigidly reserve one resource per frame to the MTs after a successful contention.

In conclusion, among MPRMA and DRAMA, the best solution is DRAMA since it guarantees a better QoS for STs and practically the same performance in terms of $E[t_{delay}]$ for WTs.

Note that a possible DRAMA refinement is obtained by separating the minislots destined to STs from those used by WTs in the access phase. According to this, due to the prioritization of ST traffic, we obtain that $P_{drop,s}$ is insensitive to the increase of WT traffic load (QoS insulation concept). Another possible refinement of the DRAMA scheme should be the insertion of access minislots distributed along the frame in order to permit more frequent and distributed access opportunities (especially if $T_f >$ RTD).

Chapter 8: Analytical methods for RRM analysis and final considerations on RRM techniques

This Chapter proposes different methods to analyze the performance of RRM techniques. The first Section deals with an approach suitable for characterizing the stability of contention-based MAC schemes; the second Section contains the analysis of a simple round robin scheme; the third Section describes an approximate method to evaluate the queuing performance in the presence of 2-MMPP sources like that described in Chapter 2 of Part II for Web surfing traffics; finally, the last Section summarizes some key aspects related to RRM strategies.

8.1   Stability study of packet access schemes

The following analysis is related to packet access schemes (MAC layer, uplink) developed for TDMA-based air interfaces (e.g., pure TDMA or even UTRA-TDD), where there is a time-basis according to which the protocol evolves.

The peculiarly of these access techniques is that they could be studied by defining a *state* that accounts for all the different possible configurations where we can found the MTs that share air interface resources. Hence, system behavior could be studied by classical methods to solve discrete-time Markov chains. Such approach cannot be practically adopted due to the state space explosion (i.e., exponentially increase in the number of permissible configurations) with the number of MTs. A simple solution, originally proposed in [109],[144],[145], is to use the *Equilibrium Point Analysis* (EPA) applied to state diagrams that model MT behaviors. For instance, we may refer to the MAC protocols described in the Chapter 7 of Part II for LEO-MSSs. Hence, a typical state diagram for the DRAMA protocol is shown in Fig. 35 for an ST (state transitions may occur at the end of each slot of the TDMA air interface). The EPA technique is based on the following definition of an *equilibrium point of the system*:

> *a point Ω in the state space is an equilibrium point if and only if it satisfies the condition that at each slot the expected change in each state variable is zero.*

The EPA approach is applied to the MT state diagram as follows:

- One state diagram has to be considered for each MT type (i.e., traffic class) that shares air interface resources.

- An equilibrium equation can be written for each state of a diagram, assuming that it is "populated" by the equilibrium number of MTs (= EPA variable).

The EPA equations for the different MT types need to be jointly solved since they are interrelated due to the use of shared resources on the air interface. Hence, an EPA system is obtained in the EPA equilibrium variables[15]. This system can be viewed as the null-gradient condition of a *potential function* V: the solution of EPA system represents an *equilibrium point* of the protocol. We can thus have both stable and unstable equilibrium points: *the access scheme behavior is acceptable only when there is a single and stable equilibrium point.* Conditions with multiple EPA solutions correspond to situations where different behaviors are possible for the protocol and, this is, of course, unacceptable. In case of a single EPA solution, the equilibrium value of a state variable is equal to the expected value of the corresponding state variable.

The advantage of the EPA approach is that it is not necessary to calculate the state transition probabilities of the discrete-time Markov chain modeling the system.

In order to illustrate the EPA method in more details, we choose to describe a simplified MAC access scheme of the Slotted Aloha type and we apply a new EPA and stability analysis, according to the seminal paper [146]. Our packet access scheme is synchronized and evolves on a slot basis, T_s, i.e., the transmission time of a packet. We have M MTs sharing the radio resources of a cell with the following random access protocol. Let us consider an MT with no packet to transmit (SIL state). It generates a new packet to be transmitted with mean rate λ pkts/slot. We assume that at most one packet can be generated in a slot by an MT. As soon as the MT generates a new packet in a slot with probability λT_s (*traffic load* of an MT), the MT

---

[15] Typically, an equation must be added for each traffic class stating that the sum of the equilibrium number of MTs in the different states equals the total number of MTs for that traffic class.

enters the contending (CON) state where it tries to transmit on a slot according to a permission probability p. Of course, $\lambda T_s \leq 1$. This attempt is successful with probability $p(1-p)^{C-1}$, where C denotes the number of MTs in the CON state. We also assume that MTs cannot generate a new packet to be transmitted until the previous packet has been successfully sent. The state diagram of an MT is shown in Fig. 38 (note that p and λT_s are two *control parameters* that influence the protocol behavior, that is the equilibrium number of MTs in the CON state).

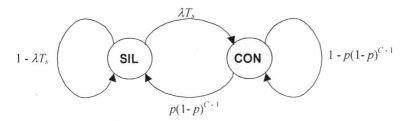

Fig. 38: MT state diagram with the Slotted Aloha-like protocol.

Let us deeply discuss the stability of this MAC scheme. A discrete-time Markov chain model with $M + 1$ states and multiple transitions from each state could be adopted to study the entire system. However, we resort here to follow the EPA approach, a particularly simple method for investigating Markov chains with too many states. Accordingly, we may write the flow balance condition at equilibrium between SIL and CON states (Fig. 38) and the normalization condition stating that the total number of MTs in SIL and CON states must be equal to M:

$$\begin{cases} s\lambda T_s = cp(1-p)^{c-1} \\ s+c = M \end{cases} \tag{51}$$

where s (c) is the equilibrium number of MTs in the SIL (CON) state.

The above system can be converted into the following equation in the unknown c (unsolvable, in a closed form) with control parameters p and λT_s and with input parameter M:

$$(M - c)\lambda T_s - cp(1 - p)^{c-1} = 0 \ . \tag{52}$$

A numerical and graphical approach has been adopted to determine the c value(s) solving (52). In Fig. 39 we diagram the surface obtained from (52), $\lambda T_s = \dfrac{cp}{(M-c)}(1-p)^{c-1}$, as a function of p and c with $M = 10$.

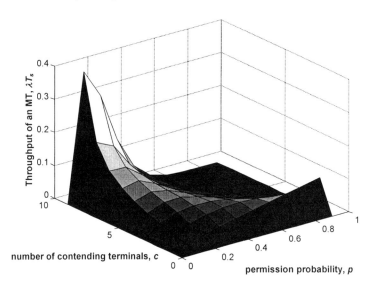

Fig. 39: Throughput values corresponding to different p and c values: this is the *catastrophe manifold* for the Slotted Aloha scheme with permission probabilities.

According to the *catastrophe theory* approach proposed in [146], (52) can be seen as the null-gradient condition of a potential, $-dV/dc = 0$. Of course, depending on the values of p, λT_s and M, the behavior of V may change so that one or more equilibrium points (i.e., EPA solutions) may occur. An equilibrium point of the potential function is stable (minimum) if $d^2V/dc^2 > 0$ and unstable (maximum) if $d^2V/dc^2 < 0$. The curve Λ in the $(p, \lambda T_s)$-plane (*control plane*) delimiting regions with different types of potential functions (= different numbers of equilibrium points) is named *bifurcation line*. It is identified as follows:

$$\begin{cases} -\dfrac{dV}{dc} = 0 \\ -\dfrac{d^2V}{dc^2} = 0 \end{cases} \Rightarrow \begin{cases} (M-c)\lambda T_s - cp(1-p)^{c-1} = 0 \\ \dfrac{d}{dc}\left[(M-c)\lambda T_s - cp(1-p)^{c-1}\right] = 0 \end{cases} . \tag{53}$$

It is easy to verify that Λ has the following expression:

$$\begin{cases} (M-c)\lambda T_s - cp(1-p)^{c-1} = 0 \\ p = 1 - e^{-\left(\frac{1}{c} + \frac{1}{M-c}\right)} \end{cases} . \tag{54}$$

In Fig. 40 we show a similar surface to that in Fig. 39 (now we consider the aggregated load $M\lambda T_s$ of the MTs in logarithmic scale, $M = 10$), but with the indication of the bifurcation line Λ (i.e., the line corresponding to which the system behavior changes in terms of the number of EPA solutions, i.e., number of equilibrium c values).

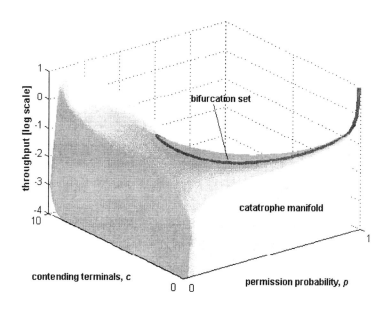

Fig. 40: Catastrophe manifold for the Slotted Aloha scheme with permission probabilities.

According to the elementary catastrophe theory [147],[148], a cusp point in Λ is characterized by (54) and the additional condition:

$$-\frac{d^3V}{dc^3} = 0 \quad \Rightarrow \quad \frac{d^2}{dc^2}\left[(M-c)\lambda T_s - cp(1-p)^{c-1}\right] = 0 \quad . \tag{55}$$

By solving (54) and (55), we obtain the cusp point as:

$$c_{cusp}(M) = \frac{M}{2} \quad , \quad p_{cusp}(M) = 1 - e^{-\frac{4}{M}} \; ,$$

$$\lambda T_s\big|_{cusp}(M) = \left(1 - e^{-\frac{4}{M}}\right) e^{-\frac{4}{M}\left(\frac{M}{2}-1\right)} \tag{56}$$

Fig. 41 shows the bifurcation line Λ for different M values in the $(p, \lambda T_s)$ *control plane*. For each M value, Ξ denotes the region in the control plane enclosed by the bifurcation line. The EPA system admits a single and stable solution when $(p, \lambda T_s) \notin \Xi$. Whereas, multiple EPA solutions occur when $(p, \lambda T_s) \in \Xi$ (we can graphically solve (52) to find that in these cases there are three possible EPA solutions). Extensive simulation runs have permitted to verify the protocol behavior in these different cases: mono-modal distribution of contending MTs with low mean value when $(p, \lambda T_s) \notin \Xi$; bi-modal distribution when $(p, \lambda T_s) \in \Xi$. The region above the upper contour of Λ (i.e., Λ^+; see Fig. 41) is characterized by high c values and, hence, by a congested contending state (the total system throughput evaluated as $\lambda T_s (M - c)$ is quite low), so that even in the presence of a single EPA solution this region cannot be adopted for typical operating conditions. The region below the lower contour of Λ (i.e., Λ^-), Y is characterized by low c values: the control parameter values must be set in this region to achieve a proper protocol behavior.

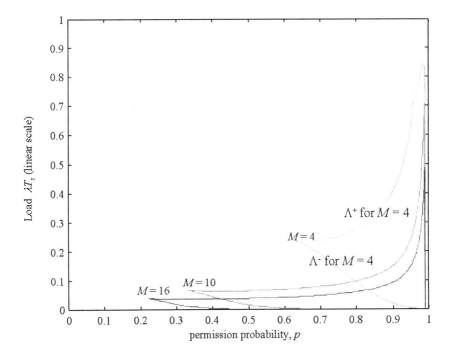

Fig. 41: Bifurcation lines and related cusp points for different M values. Note that the EPA system admits three solutions in the region enclosed by the cusp.

Let us assume that we have a given λT_s value (admissible by the protocol, i.e., $\lambda T_s \leq 1$, but also allowing a non-congested behavior, i.e., $\lambda T_s \leq \lambda T_s \mid_{cusp}(M)$), $\overline{\lambda T_s}$, so as to operate in Y. The *total system throughput* is represented by the flow $(M - c)\lambda T_s$, where c satisfies (52).

The definition in the control plane of the area Y where the protocol allows a good behavior permits to proceed further for the identification of the permission probability value that maximizes the total system throughput for a given $\overline{\lambda T_s}$ value. This is equivalent to select the p value that minimizes the EPA solution c for a given $\overline{\lambda T_s}$ value. For the sake of conciseness, we limit ourselves in stating that such an optimum p value is reached for the limiting p value at the intersection of the load $\overline{\lambda T_s}$ line and the bifurcation line. However, a safety margin must be included to avoid the design of the system just on the bifurcation line. According to this method, the selection of the optimum p value is thus adaptive to the load $\overline{\lambda T_s}$ value.

In conclusion, the EPA analysis is a powerful tool that permits a clear understanding of the behavior of a packet access scheme and also allows the optimized selection of parameters.

8.2   Analysis of Round Robin traffic scheduling

Let us consider the management of Poisson arrivals of messages with a truncated Pareto length distribution, as in (6) of Chapter 2. In this study, we refer to the GPRS air interface resources described in Chapter 3 of Part II. The following analysis considers a single PDCH slot per frame and is embedded at the end of a multi-frame period; we name *packet* the information content carried out by a block. We assume that there is a pre-emptive priority conversational traffic (produced by transactional applications) that uses radio blocks with probability r that is multiframe-to-multiframe independent (Bernoulli traffic). Hence, a GPRS radio block is available to carry the transmission of messages with probability $1 - r$.

As shown in the Introduction of Part II, the ideal *Processor Sharing* (PS) scheme (i.e., a bit-by-bit round robin among all the users with messages in the queue) is more convenient than the FIFO discipline to manage the transmissions of messages with heavy-tailed length distributions. Such condition is verified by the message length distribution in (6). A very simple PS implementation (for packets of the same length and equal priority) is given by the *Round Robin* (RR) approach (see Fig. 10). In particular, we refer here to two distinct cyclic service mechanisms for the messages of different users in the downlink transmission queue of a PDCH: (*i*) one entire message is transmitted per user per cycle; (*ii*) one packet/block is transmitted per user per cycle. Let T_{pkt} denote the packet transmission time (i.e., one multi-frame, under our assumptions).

Let $T(z)$ denote the *Probability-Generating Function* (PGF) [5] of the block transmission time (geometrically distributed) in multiframes:

$$T(z) = \frac{(1-r)z}{1-rz} \quad . \tag{57}$$

We limit the following analysis of the RR scheme to the simple case of geometrically distributed message lengths as in (21), showing that also in these circumstances the RR approach yields some benefits with respect to the FIFO one. Let λ_d denote the mean message arrival rate for the Poisson process.

The queue for PDCH transmissions can be split into M_d *virtual* sub-queues of the $M/G/1$ type (one for each user) that are cyclically served. We refer to a tagged sub-queue and we embed this study to the end of message transmission. Each $M/G/1$ sub-queue is served only if it contains data to be transmitted, otherwise the service is instantaneously given to the next queue in the list. In this study we neglect the differentiation in the transmission time that occurs when an arriving message finds an empty queue (conservative analysis).

Let P_0 denote the empty sub-queue probability. The number of sub-queues served (one message or one packet transmitted at most from each of them per cycle) to complete the cycle is binomially distributed from 0 to $M_d - 1$ with parameter P_0; the corresponding PGF, $B(z)$, is:

$$B(z) = [P_0 + z(1 - P_0)]^{M_d - 1} \quad . \tag{58}$$

Let $L_d(z)$ denote the PGF of the message length l_d in packets:

$$L_d(z) = \frac{\dfrac{z}{E[l_d]}}{1 - \left\{1 - \dfrac{1}{E[l_d]}\right\}z} \quad . \tag{59}$$

In the two RR schemes, the transmission time of a message is: (*i*) sum of the effective message transmission time and the time to complete the cycle (= service for the remaining $M_d - 1$ queues; one message for each, if any); (*ii*) the packet transmission time and a time to complete the cycle, both summed for all the packets in the message. All these contributions are assumed to be statistically independent (this is similar to the independence assumption for a network of queues [5]). Hence, the message transmission time has the following PGFs: $L_d(T(z))B[L_d(T(z))]$ in the message-based RR scheme and $L_d[T(z)B(T(z))]$ in the packet/block-based RR scheme. Moreover, the PGF of the number of message arrivals from a user in T_{pkt} is $A_{u,T_{pkt}}(z) = e^{\lambda_d T_{pkt}(z-1)}$. Hence, the PGF of the number of message arrivals in the transmission time of a message, $A_{RR}(z)$, results as:

$$A_{RR}(z) = \begin{cases} L_d\left\{T\left(A_{u,T_{pkt}}(z)\right)\right\}B\left\{L_d\left[T\left(A_{u,T_{pkt}}(z)\right)\right]\right\}, & \text{message-based RR} \\ L_d\left\{T\left(A_{u,T_{pkt}}(z)\right)B\left[T\left(A_{u,T_{pkt}}(z)\right)\right]\right\}, & \text{packet-based RR} \end{cases} \quad (60)$$

We derive P_0 by imposing $P_0 = 1 - A_{RR}'(z = 1)$. In both cases, we have:

$$P_0 = \frac{1 - r - \lambda_d T_{pkt} E[l_d] M_d}{1 - r - \lambda_d T_{pkt} E[l_d](M_d - 1)} \quad . \quad (61)$$

We consider the classical $M/G/1$ expression of the PGF of the number of messages in each sub-queue, $P_{RR}(z)$, and we adopt the transformation $z = 1 - s/\lambda_d$ that yields the *Laplace-Stieltjes Transform* (LST) of the pdf of the message delay, $D_{RR}(s)$ [5]:

$$D_{RR}(s) = P_{RR}(z = 1 - s/\lambda_d) = P_0 \frac{(z-1)A_{RR}(z)}{z - A_{RR}(z)}\Big|_{z=1-s/\lambda_d} \quad . \quad (62)$$

In the FIFO case, the LST of the pdf of the message delay, $D_{FIFO}(s)$, is obtained by considering the global queue with an $M/G/1$ model, as follows:

$$D_{FIFO}(s) = P_{RR}(z = 1 - s/(M_d\lambda_d)) =$$
$$= \left(\frac{1 - r - \lambda_d T_{pkt} E[l_d] M_d}{1 - r}\right) \frac{(z-1)A_{FIFO}(z)}{z - A_{FIFO}(z)}\Big|_{z=1-s/(M_d\lambda_d)} \quad (63)$$

In order to obtain numerically the pdfs that correspond to $D_{RR}(s)$ and $D_{FIFO}(s)$, we use the change of variable $s = j2\pi f$, where j is the imaginary unit ($j^2 = -1$). Then, we sample in the frequency domain both $D_{RR}(f_n) = D_{RR}(s = j2\pi f_n)$ and $D_{FIFO}(f_n) = D_{FIFO}(s = j2\pi f_n)$ with interval f_c (see below) and we apply the *Inverse Fast Fourier*

Transform (IFFT) algorithm, by considering scaled versions by $1/T_c$, where T_c is the sampling interval in the time domain. We make the approximation that the Fourier components are negligible for $f > f_{max}$. We determine f_{max} so that $T_c = 1/(2f_{max})$ is the smallest possible delay value (i.e., T_{pkt}). Moreover, we expect that the pdf vanishes to zero for high t values. Hence, the number of samples of the IFFT, N, can be determined by considering the approximation that the pdf is equal to zero for $t > NT_c$. Since the pdfs are unknown, we will use suitable values for NT_c and we will *a posteriori* verify that the obtained pdfs are negligible for $t > NT_c$. Note that N represents also the number of samples in the frequency domain in $2f_{max}$, so that f_c is determined as well. The cumulative distribution functions (cdfs) have been obtained through numerical integration.

Fig. 42 compares the cdfs of the message delay obtained from analysis ($f_{max} = 25$ Hz, $NT_c = 10.24$ s) and simulations for the two RR schemes in the GPRS CS-4 case with Poisson arrivals of the messages (mean arrival rate of 1.52 msg/s, mean message length of 481 bytes/msg), 1 PDCH per frame, $r = 0.1$ Erlang and $M_d = 2$.

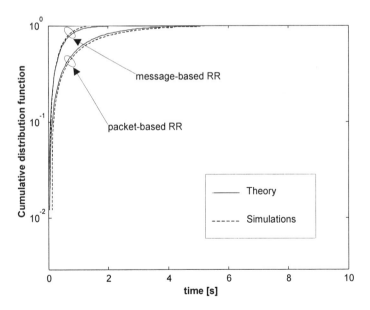

Fig. 42: Comparison of the cdfs of the message transmission delay obtained from both analysis and simulations in the GPRS case with 1 PDCH/frame, $r = 0.1$ Erlang, CS-4 coding, $M_d = 2$ users.

Fig. 42 highlights a satisfactory agreement between theoretical and simulation results. We note that, even if the block-based RR scheme guarantees a maximum delay value for block transmissions (thus allowing to meet minimum bit-rate requirements), the message-based RR scheme achieves much better results for 90-th and 95-th percentiles of the message delay, thus improving the QoS perceived by users.

8.3 2-MMPP traffic delay analysis

We consider here the delay analysis for the case of the Web traffic source described in Chapter 2 of Part II. We assume that it is possible to transmit packets on a time slot basis ($T_{pkt} = T_s$). Each datagram is fragmented in packets before transmission. We have adopted the following model to study the system: $2\text{-}MMPP^{[P]}/D/1$, where $2\text{-}MMPP^{[P]}$ stands for the 2-MMPP arrival process of datagrams with truncated Pareto distribution of a Web source, D is the deterministic packet service time, "1" means that only one packet can be transmitted per slot.

We start by deriving the mean packet delay, $T_{d,pkt}$. We embed the model at the end of slot instants. In what follows, we have modified the approximated analytical approach proposed in [149] by taking into account that packets have a compound arrival process due to both the two-state generation process of datagrams and their variable length in packets. Hence, the packet arrival process is characterized by the following PGF matrix:

$$Q(z) = \begin{bmatrix} A_{11}(z) & A_{12}(z) \\ A_{21}(z) & A_{22}(z) \end{bmatrix} = \begin{bmatrix} p_{11}e^{\lambda_p T_s [L_w(z)-1]} & p_{12} \\ p_{21}e^{\lambda_p T_s [L_w(z)-1]} & p_{22} \end{bmatrix} \quad (64)$$

where:

- each $A_{ij}(z)$ denotes the *Probability-Generating Function* (PGF) of the number of packets arrived in a slot where the Web source makes the transition from state i to state j ($i, j \in \{1, 2\}$: value 1 is related to the packet call state and value 2 is related to the reading time state);

- $\lambda_p = 2q$ is the mean datagram arrival rate in the packet call state;

- $L_w(z)$ is the PGF of distribution of the datagram length l_w in packets shown in (6);

- $p_{12} = 1 - e^{-T_s/m_{Lpc}}$ is the probability that the source leaves the packet call state in T_s;

- $p_{11} = 1 - p_{12}$;

- $p_{21} = 1 - e^{-T_s/m_{Dpc}}$ is the probability that the source leaves the reading state in T_s;

- $p_{22} = 1 - p_{21}$.

Let $\boldsymbol{s} = (s_1, s_2)^T$ denote the state probability vector for the Markov modulating process of the 2-MMPP source (apex T denotes the transpose vector); in particular, s_1 is the probability of the packet call state and s_2 is the probability of the reading time state. Vector $\boldsymbol{s}$ can be obtained as:

$$s_1 = \frac{p_{21}}{p_{12} + p_{21}} \equiv \psi_w \quad , \quad s_2 = \frac{p_{12}}{p_{12} + p_{21}} . \tag{65}$$

We embed the system at the end of each slot and each source transmits one packet per slot. According to [149] and on the basis of (64), we obtain the following expression for the mean packet delay, $T_{d,pkt}$:

$$T_{d,pkt} = 1 + \frac{\lambda_1''(1)}{2\rho_w[1-\rho_w]} + \frac{\xi_1'(1)}{\rho_w} \quad \text{[slots]} \tag{66}$$

where ρ_w is the traffic intensity generated by a Web traffic source (7) and where parameters $\lambda_1''(1)$ and $\xi_1'(1)$ have quite complex definitions (see [58]) that lead to the following expressions:

- $\lambda_1'(1) = \lambda_w T_s E[l_w] \equiv \rho_w$;

- $\lambda_1''(1) = \lambda_1'(1)\left[\dfrac{E[l_w^2]}{E[l_w]} - 1 + \lambda_p T_s E[l_w]\right] + 2\dfrac{\rho_w^2}{p_{12} + p_{21}}$;

- $\xi_1'(1) \cong \{s^T Q(0) u_1'(1)\} / \{s^T Q(0) I\}$, being $u_1(1)$ the first eigenvector of

  matrix $Q(z)$ with related eigenvalue $\lambda_1(z)$: $u_1'(1) = \dfrac{\lambda_2'(1)}{p_{12}} (s_2, -s_1)^T$

  and $I = (1, 1)^T$.

Note that this analysis can be easily extended to the case where there are M_w Web traffic sources that share the same transmission queue by adapting $\lambda_1''(1)$, $\xi_1'(1)$ and $T_{d,pkt}$ definitions [149].

Finally, the mean datagram transmission delay, $E[t_{delay}]$, can be obtained by considering that the mean packet delay is related to the transmission of one packet in the middle of a datagram and, therefore, the complete mean datagram delay requires to account also for the transmission of the remaining half datagram:

$$E[t_{delay}] = \left(T_{d,pkt} + \frac{E[l_w]}{2} \right) T_s \quad [s] \tag{67}$$

The graph in Fig. 43 compares simulation and analytical results for the mean packet delay $T_{d,pkt}$ for a case with the Web traffic model with $q = 8$, a packet payload of 48 bytes (like WATM) and a packet transmission time $T_s = 4.32 \times 10^{-5}$ s. We may note that theory slightly overestimates the mean packet delay, but approximations reduce with the traffic load.

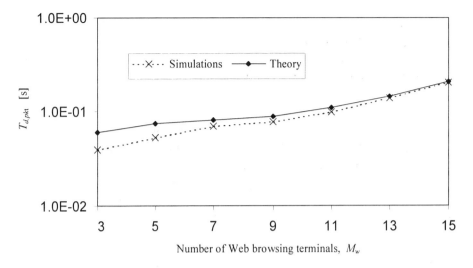

Fig. 43: Comparison between theoretical and analytical results for the delay analysis with 2-MMPP Web traffic sources.

8.4   Lessons learned on RRM strategies

We conclude our study on RRM schemes proposing some considerations that can be useful when designing protocols to share radio resources among distinct traffic classes with different QoS requirements. In particular, we focus below on scheduling and priority aspects.

1. A centralized scheduler at the base station may allow the centralized management of different traffic classes on both uplink and downlink. In particular, distinct transmission queues must be used, one for each traffic class in order to manage better the different QoS requirements. A scheduler and a traffic regulator must decide on a regular basis the radio resources destined for the transmissions from the different queues. Scheduling decisions must be based on requests coming from the different (uplink and downlink) terminals and have to fulfill QoS requirements of the traffic classes; a fair resource sharing is needed within a class.

2. RR scheduling allows good percentile delay performance and a fair sharing of resources among the sources of a given traffic class. The

RR approach also guarantees minimum bit-rate values for the managed traffic sources so that it is particularly suited to support the traffics generated by Web traffic sources and regulated by the TCP protocol (interactive traffic class).

3. The integration of RR (polling) with a token bucket policer for downlink (uplink) transmissions allows the realization of a centralized scheduler that guarantees a fair sharing of resources within a traffic class and accounts for prioritization among different traffic classes. Such approach is particularly suited for real-time (i.e., conversational and streaming) traffics and interactive traffics. An adaptive RR scheme could avoid the transmission of non-urgent traffics when they experience unfavorable radio propagation conditions.

4. RR schemes must preferably service one higher-layer message at one, in order to improve the percentile delay performance of the RRM scheme.

5. Priority must be given to traffic classes as follows (referring to 3G systems classification):

   conversational and *streaming* > *interactive* > *background.*

   Moreover, RRM has to provide QoS insulation to the different traffic classes with exceptions when interactive traffics experience starvation (that may reduce the throughput of sources regulated by TCP).

6. Elastic capacity (i.e., not fixed) has to be provided to bursty traffic sources with heavy-tailed message length in order to reduce transmission delays: the scheduler must be able to react quickly to the generation of such traffics by assigning them temporarily higher capacity.

7. For CDMA air interfaces, RRM means not only the scheduling of traffics, but also the definition of physical layer parameters such as transmission power level and processing gain value. We have shown that messages that occupy the resources for the shortest time (i.e., having low processing gain values) must be prioritized to reduce the service delays.

8. TDD air interfaces are more suited to support multimedia and asymmetric traffics, since resources can be dynamically shared between downlink and uplink on the basis of the related traffic loads. Due to the reciprocal nature of TDD links, the physical parameters (i.e., power) to be used in uplink can be estimated from the received downlink transmission levels from the base stations, thus simplifying the tasks of power control algorithms (open loop schemes). TDD air interfaces, therefore, can be adopted for the realization of WLAN to be integrated in 4G systems (e.g., HIPERLAN/2 case).

9. Access protocols (uplink) that are based on random transmissions of request packets on shared resources must be carefully designed to avoid that they experience congestion (with consequent high number of collisions among attempts).

   - First of all, traffic sources that generate a continuous stream of traffic should use piggybacked requests in the transmitted packets in order to update their transmission needs, thus avoiding random access attempts.

   - Information resources should be taken distinct from those used in the access phase.

   - The parameters of the access scheme must be adaptively defined (depending on traffic load and radio channel conditions) in order to improve the throughput of successful requests.

   - Finally, we have shown that the EPA approach and the related stability analysis can be employed to optimize the design of access protocols.

Chapter 9: A first solution towards the mobile Internet: the WAP protocol

This Chapter deals with the analysis of the techniques that allow a mobile fruition of the Internet. In particular, we focus here on the *Wireless Application Protocol* (WAP). Moreover, the next Chapter will consider the TCP/IP protocol suite and its behavior in the presence of wireless links.

WAP is the result of the WAP Forum's efforts for allowing mobile users to access easily the Internet. WAP contains a lightweight protocol stack (based on the TCP/IP one) well suited for the wireless scenario. It also includes an application environment for the development of user applications for mobile phones. A review of the protocol stack and system architecture is provided here together with some aspects related to the tools to develop WAP applications.

9.1   Introduction to WAP

It is expected that within few years the number of mobile devices accessing the Internet will exceed the number of *Personal Computers* (PC). The use of mobile terminals is becoming attractive, since they allow the access on the move and require a shorter set-up time than the initial power on of a PC. However, Internet services have not been developed for mobile devices, they are not suited to small displays, they are not personalized or location-dependent. A first answer to these needs is represented by the *Wireless Application Protocol* (WAP). WAP is an open, global specification that empowers mobile users with wireless devices to access easily the Internet [150],[151].

WAP is the result of the WAP Forum's efforts to promote industry-wide specifications for developing applications and services that operate over wireless communication networks [153]. WAP Forum was formed after a US network operator, Omnipoint, issued a tender for the supply of mobile information services in early 1997. It received several responses from different suppliers using proprietary techniques for delivering the information such as *Smart Messaging* from Nokia and *Handheld Device Markup Language* (HDML) from Phone.com.

Omnipoint informed the tender responders that it would not accept a proprietary approach and recommended that various vendors got together to explore the definition of a common standard. After all, there was not a great deal of difference between the different approaches, which could be combined and extended to form a powerful standard. These events triggered the development of WAP, with Ericsson and Motorola joining Nokia and Phone.com as the founder members of the WAP Forum. At present, WAP Forum encompasses more that 500 members.

WAP is the *de facto* standard for porting Internet services to wireless devices, such as mobile phones and *Personal Digital Assistants* (PDA). WAP can be built on many mobile phone operating systems, including PalmOS, EPOC, Windows CE and JavaOS [152]. WAP contains also a micro-browser according to which the information received is interpreted in the handset and presented to the user. It is designed to work with most wireless networks such as CDPD, CDMA, GSM, PDC, PHS, TETRA and DECT [152]. WAP devices, despite the current rather limited user interface, provide a valuable means to access corporate and public services.

9.2   WAP architecture

The WAP network architecture envisages both *WAP servers*, hosting pages designed in a suitable markup language and *WAP gateways* between the wireless network and the wireline Internet [153]. WAP 1.0 was based on *Wireless Markup Language* (WML), a subset of the *eXtensible Markup Language* (XML). The basic markup language in WAP 2.0, namely WML2, is based on the *eXtensible HyperText Markup Language* (XHTML), as defined by the *World Wide Web Consortium* (W3C) [154]. By using the XHTML modularization approach, the WML2 language is very extensible, permitting additional language elements to be added as needed.

The WAP protocol architecture is based on a client/server model (Fig. 44). The client Web browser makes a request for a Web page. This request is sent to the WAP proxy that acts a gateway to the Internet [155]. Through a protocol conversion a *HyperText Transfer Protocol* (HTTP) request is thus sent to the appropriate Web server. The response is a byte stream of ASCII text, which is a *HyperText Markup Language* (HTML) Web page. The use of *Common Gateway Interface* (CGI) programs or Java Servlets allows for the dynamic creation of HTML pages using content stored in a database.

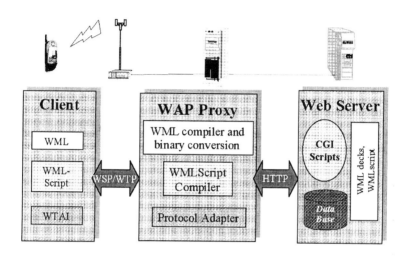

Fig. 44: Interoperation of WAP elements.

The Web page is sent to the WAP proxy that firstly translates it from HTML to WML (Fig. 45, step 2). Finally, a page conversion is performed in a compact binary representation that is suitable for wireless networks (Fig. 45, step 3) [156]. The mobile client can directly receive WML pages from WAP servers (hosting pages in the WML format) in the mobile network, without the involvement of the WAP gateway.

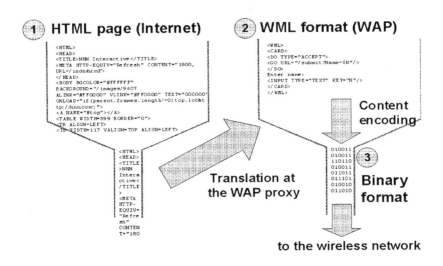

Fig. 45: WML compiling and encoding process.

Although reusing of existing Internet content by means of on-the-fly adaptations and translations is an explicit goal, test realizations of WAP gateway and proxy servers show that the creation of new contents (explicitly designed in WML) is a more effective option.

Since a mobile user cannot use a QWERTY keyboard or a mouse, WML documents are structured into a set of well-defined units of user interactions called *cards*. Each card may contain instructions for gathering user input, information to be presented to the user, etc. A single collection of cards is called *deck*, which is the unit of content transmission, identified by a *Uniform Resource Locator* (URL) [157]. After browsing a deck, the WAP-enabled phone displays the first card; then, the user decides whether to proceed or not to the next card of the same deck. WML content is scalable from a two-line text display on a basic device to a full graphic screen on the latest smart phones and communicators. WML supports:

- Text (bold, italics, underlined, line breaks, tables);
- Black and white images (wireless bitmap format, WBMP);
- User input;
- Variables;

- Navigation and history stack;

- Scripting (WMLScript), a lightweight scripting language, similar to JavaScript.

In particular, WML includes support for managing user agent state by means of variables and for tracking the history of the interaction. Moreover, WMLScripts are sent separated from decks and are used to enhance the client *Man-Machine Interface* (MMI) with sophisticated device and peripheral interactions.

9.3 WAP protocol stack

WAP protocol and its functions are layered similarly to the *OSI Reference Model* [158]. In particular, the WAP protocol stack is analogous to the Internet one. The WAP protocol layers at the client, at the gateway and at the Web server are detailed in Fig. 46. Each layer is accessible by the layers above, as well as by other services and applications. The WAP layered architecture enables other services and applications to utilize the features of the WAP stack through a set of well-defined interfaces. A brief survey of the protocols at the different WAP layers is provided below.

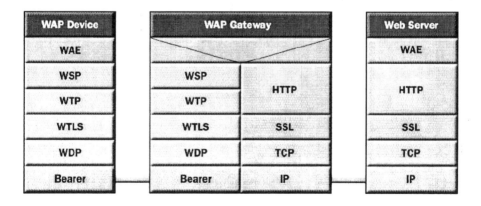

Fig. 46: WAP 1.0 protocol architecture at different interfaces.

Wireless Application Environment **(WAE)**: WAE characterizes an application framework for wireless devices such as mobile phones, pagers and PDAs. WAE specifies the markup languages and acts as a container for applications such as a micro-browser. In particular, WAE encompasses the following parts:

- WML micro-browser;

- WMLScript virtual machine;

- WMLScript standard library;

- *Wireless Telephony Application Interface* (i.e., telephony services and programming interfaces);

- WAP Content Types.

The two most important formats defined in WAE are WML and WMLScript formats. A WML encoder at the WAP gateway, or "tokenizer", converts a WML deck into its binary format (Fig. 45, step 3) [156] and a WMLScript compiler transforms a script into byte-code. This process allows a significant compression of the data to be sent, for an efficient use of air interface resources.

Wireless Session Protocol **(WSP)**: WSP provides the application layer of WAP (i.e., WAE) with a consistent interface for two-session services. The first is a connection-oriented service above the *Wireless Transaction Protocol* (WTP). The second is a connectionless service operating above a secure or non-secure datagram service (*Wireless Datagram Protocol*, WDP). WSP is the equivalent of the HTTP protocol in both the Internet and WAP 2.0 release that supports the TCP/IP levels in the protocol stack.

Wireless Transaction Protocol **(WTP)**: WTP runs on top of a datagram service (such as *User Datagram Protocol*, UDP) and provides a lightweight transaction-oriented protocol that is suitable for implementation in mobile terminals. WTP provides three classes of transaction services: unreliable one-way request, reliable one-way request and reliable two-way request-response.

Wireless Transport Layer Security **(WTLS)**: WTLS is a security protocol based upon the industry-standard *Transport Layer Security* (TLS) protocol. WTLS is intended for use with the WAP transport protocols and has been optimized for wireless communication networks. It includes data integrity checks, privacy on the WAP gateway-to-client leg and authentication.

***Wireless Datagram Protocol* (WDP)**: WDP is transport layer protocol in WAP [159]. WDP supports connectionless reliable transport and bearer independence. WDP offers consistent services to the upper layer protocols of WAP and operates above the data capable bearer services supported by various air interfaces. Since WDP provides a common interface to upper-layer protocols, security, session and application layers are able to operate independently of the underlying wireless network. At the mobile terminal, the WDP protocol consists of the common WDP elements plus an adaptation layer that is specific for the adopted air interface bearer. The WDP specification lists the bearers that are supported and the techniques used to allow WAP protocols to operate over each of them [152]. The WDP protocol is based on UDP. UDP provides port-based addressing and IP provides *Segmentation And Re-assembly* (SAR) in a connectionless datagram service. When the IP protocol is available over the bearer service, the WDP datagram service offered for that bearer will be UDP.

9.3.1 Bearers for WAP on the air interface

Let us refer to the Global System for Mobile communications (GSM) network, where the following bearer services can be adopted to support WAP traffic [118]:

- *Unstructured Supplementary Services Data* (USSD);

- *circuit-switched Traffic CHannel* (TCH);

- *Short Message Service* (SMS);

- *General Packet Radio Service* (GPRS), plain data traffic;

- *Multimedia Messaging Service* (MMS) over GPRS.

Let us compare these different options. TCH has the disadvantage of a 30-40 s connection delay between the WAP client and the gateway, thus making it less suitable for mobile subscribers.

Both SMS and USSD are inexpensive bearers for WAP data with respect to TCH, leaving the mobile device free for voice calls. SMS and USSD are transported by the same air interface channels. SMS is a store-and-forward service that relies on a *Short Message Service Center* (SMSC). Whereas, USSD is a connection-oriented (no store-and-forward) service, where the *Home Location Register* (HLR) of the GSM network receives/routes messages from/to the users. The SMS bearer is well suited for WAP push applications (available from WAP release 1.2), where the user is automatically notified each time an event occurs. USSD is particularly useful for supporting transactions over WAP.

Finally, GPRS radio transmissions allow a high capacity (up to 170 kbit/s using all the slots of a GSM carrier with the CS-4 coding scheme) that is shared among mobile phones according to a packet switching scheme. Hence, GPRS can provide a powerful scheme for WAP contents delivery.

9.4   Tools and applications for WAP

The WAP programming model is similar to the WWW programming one. This fact provides several benefits to the application developer community, including a proven architecture and the ability to leverage existing tools (e.g., Web servers, XML tools, etc). Optimizations and extensions have been made in order to match the characteristics of the wireless environment. Different WAP browsers can be found in reference [160]; they are useful tools for developing WAP-based services for mobile users. WAP allows customers to easily reply to incoming information on the phone by adopting new menus to access mobile services.

Existing mobile operators have added WAP support to their offering, either by developing their own WAP interface or, more usually, partnering with one of the WAP gateway suppliers. WAP has also given new opportunities to allow the mobile distribution of existing information contents. For example, CNN and Nokia teamed up to offer CNN Mobile. Moreover, Reuters and Ericsson teamed up to provide Reuters Wireless Services.

New mobile applications that can be made available through a WAP interface include:

- Location-aware services;

- Web browsing;

- Remote LAN access;

- Corporate e-mail;

- Document sharing / collaborative working;

- Customer service;

- Remote monitoring such as meter reading;

- Job dispatch;

- Remote point of sale;

- File transfer;

- Home automation;

- Home banking and trading on line.

Another group of important applications is based on the WAP push service that allows contents to be sent or "pushed" to devices by server-based applications via a push proxy. Push functionality is especially relevant for real-time applications that send notifications to their users, such as messaging, stock price and traffic update alerts. Without the push functionality, these applications would require the devices to poll application servers for new information or status. In cellular networks such polling activities would cause an inefficient and wasteful use of the resources. WAP push functionality provides control over the lifetime of pushed messages, store-and-forward capabilities at the push proxy and control over the bearer choice for delivery.

Interesting WAP applications are made possible by the creation of dynamic WAP pages by means of the following different options:

- Microsoft ASP;

- Java and Servlets or *Java Server Pages* (JSPs) for generating WAP decks;

- *XSL Transformation* (XSLT) for generating WAP pages adapted for displays of different characteristics and sizes.

Alternative approaches to the use of WAP for mobile applications could be as follows:

- *Subscriber Identity Module* (SIM) - *Toolkit*: the use of SIMs or smart cards in wireless devices is already widespread.

- *Windows CE*: this is a multitasking, multithreaded operating system from Microsoft designed for including or embedding mobile and other space-constrained devices.

- *JavaPhoneTM*: Sun Microsystems is developing PersonalJavaTM and a JavaPhoneTM *Application Programming Interface* (API), which is embedded in a JavaTM virtual machine on the handset. Thus, cellular phones can download extra features and functions from the Internet.

SIM Toolkit and Windows CE are present days technologies as well as WAP. SIM Toolkit implies the definition of a set of services "embedded" on the SIM that allow contacting several service provides through the mobile phone network. The Windows CE solution is based on an operating system developed for mobile devices, supporting different applications. Finally, JavaPhoneTM will be the most sophisticated option for the development of device-independent applications.

Within ETSI and 3GPP, activities are in progress for the definition of new architectures providing mobile information services. Accordingly, a new standard, called *Mobile station application Execution Environment* (MExE), has been defined [161]. MexE is a VHE technology, according to the description given in Chapter 3 (Section 3.6) in Part I. In particular, in order to insure the portability of a variety of applications, across a broad spectrum of multi-vendor mobile terminals, a dynamic and open architecture has been conceived in MExE for both the *Mobile Station* (MS) and the SIM, i.e., a common set of APIs and development tools. MExE is based on the idea to specify a terminal-independent execution environment on the client device (i.e., MS and SIM) for non-standardized applications and to implement a mechanism that allows the negotiation of supported capabilities (taking into account available bandwidth, display size, processor speed, memory, MMI). The key concept of the MExE service environment to make mobile-aware applications (i.e., aware of MS capabilities, network bearer characteristics and user preferences) is the introduction of MExE classmarks that have been standardized as follows:

-   *MExE classmark 1*: it is based on WAP, requires limited input and output facilities (e.g., as simple as a 3 lines by 15 characters display and a numeric keypad) on the client side and is designed to provide quick and cheap information access even over narrow and slow data connections.

-   *MExE classmark 2*: it is based on PersonalJavaTM, provides and utilizes a run-time system requiring more processing, storage, display and network resources, but allows powerful applications and more flexible MMIs. MExE classmark 2 also includes the support for MExE classmark 1 applications (via the WML browser).

Chapter 10: The mobile Internet

Recent years have seen a strong development of wireless and mobile devices, such as palmtops, personal communicators and *Personal Digital Assistants* (PDA), characterized by increasing processing capabilities and memory storage. Such devices give the possibility of accessing the network, sending and receiving e-mails and browsing the Web while on the move. The wish to connect to the Internet and maintain communications anytime and anywhere has led to the need of the mobile Internet. Today, support of Internet services in a mobile environment is an emerging requirement. The issues to be faced in order to support the wireless and mobile Internet are related to different protocol layers:

- *Network layer*: the *Internet Protocol* (IP) needs modifications in order to manage the routing to/from a mobile node;

- *Transport layer*: the *Transmission Control Protocol* (TCP) should be refined in order to work efficiently on error-prone wireless links.

10.1   IP and mobility

The TCP/IP suite was originally designed to work with wired networks. One basic problem with mobile Internet is related to the routing mechanism for delivering packets to mobile stations. As a matter of fact, IP addresses are defined according to a topological relation with the connected nodes, assuming that any node has always the same point of attachment to the Internet. According to the original IP addressing scheme, when a computer moves to a new point of attachment, it should be assigned a new IP configuration (i.e., IP address, netmask and default router) in order to be visible in the Internet. In the scenario depicted in Fig. 47, datagrams addressed to the laptop in subnet B will be always routed through link B; if this node moves to subnet C, it will not receive datagrams anymore, because packets will still be routed to link B.

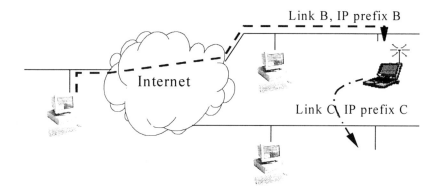

Link B, IP prefix B

Internet

Link C, IP prefix C

Fig. 47: The problem of the movement of an IP node.

Mobile IP [162] was introduced by IETF with the purpose to support mobile devices while dynamically changing their access points to the Internet.

The mobility concept can be categorized in two classes [163]:

- *Macro-mobility*: this term relates to movements of a mobile node among different *IP domains* or different wireless access networks; mobility management is held by a macro-mobility scheme, named Mobile IP.

- *Micro-mobility*: it relates to movements carried out among different micro-cells within the same IP domain. Mobile IP is not appropriate to support fast, seamless handoffs between cells and a micro-mobility scheme is needed for managing micro-mobility.

10.1.1 Mobile IP

Both ends of a TCP session (connection) need to keep the same IP address for the whole life of the session. This address, assigned for an extended period of time to a mobile node, is called *home address* and it remains unchanged regardless of where the node is attached to the Internet. As explained before, the IP address needs to be changed when a network node moves to a new place in the network. This new address, called *care-of-address*, is associated to the mobile node while it is away from home and it is used for routing purposes. Mobile IP solves the IP mobility problem by means of a routing approach, managing a dynamic association between a care-of-address to a home address, called a *binding*.

According to this mechanism, Mobile IP is an extension to IP protocol, allowing a mobile node to use two different IP addresses, a *static* one (home address) for its identification and a *dynamic* one (care-of-address) for routing. In such a way the node can continue receiving datagrams, independently of its location.

The Mobile IP Working Group has developed routing support to permit IP nodes (routers and hosts) using either IPv4 or IPv6 to seamlessly roam among IP sub-networks. It allows *macro-mobility* management independent of radio access technology and provides seamless roaming among heterogeneous wireless networks (i.e., GPRS, UMTS and wireless LAN). Transparency above the IP layer is supported, including the maintenance of active TCP connections and UDP port bindings. The cellular and wireless industry is considering using Mobile IP as a technique for IP mobility for wireless data.

10.1.2 Micro-mobility and the Cellular IP approach

Even if Mobile IP provides a simple and scalable mobility scheme, it is not appropriate for high mobility and seamless handoffs. In fact, it envisages that every time a node migrates, a local address must be obtained and communicated to a distant location directory, called *home agent*. This updating procedure, together with route optimization, introduces delays and data transfer disruption while the correspondent node obtains the new binding. The effect of these delays grows with the frequency of handoffs. Moreover, when host mobility becomes ubiquitous and cell size smaller, the traffic load generated by the update messages can have a drastic effect on the Internet and on the home agent as well, being proportional to the number of mobile hosts.

Cellular IP [164],[165] is one of the most attracting schemes for managing micro-mobility. It is aimed at optimizing handoffs in a restricted geographical area, rather than supporting global mobility. Fig. 48 depicts a possible scenario in which local and wide area mobility are separated: Mobile IP manages global mobility, Cellular IP manages migrations at the local level (i.e., within the wireless access network).

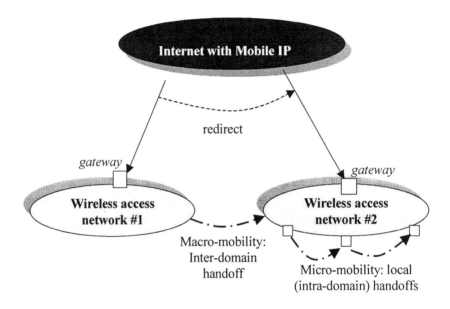

Fig. 48: Global and local mobility.

According to this general scenario, handoffs within the access network are locally handled. Hence, handoffs can be faster and the impact on active data sessions is limited.

Cellular IP defines a wireless access network architecture and protocol for managing micro-mobility. It is based on cellular technology principles for mobility management, passive connectivity (i.e., paging) and handoff support. It operates at the network layer, substituting the IP routing mechanism in the wireless access network, without modifying the packet format and the IP forwarding mechanism.

The Cellular IP *node* embeds different functions, such as: wireless access point, IP packet routing and cellular control functionality, traditionally found in MSC and BSC. The nodes implement Cellular IP integrated routing and location management and are built on regular IP forwarding engine.

A gateway connects the Cellular IP network to the Internet. Its IP address is used by mobile hosts attached to the network as their Mobile IP care-of address (see Fig. 49).

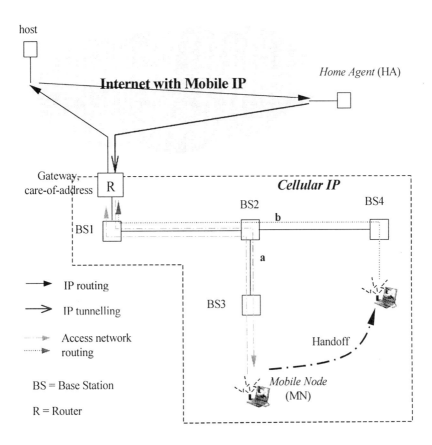

Fig. 49: Cellular IP handoff scenario and routing.

Uplink routing (i.e., from MN to gateway) is performed on a hop-by-hop basis. Nodes on the route cache the path taken by uplink packets. After MN data transmissions (see Fig. 49), the *routing cache* in BS2 includes a mapping (MN, a), indicating that MN is reachable through interface "a" (see the path labeled with "b" in Fig. 49). Cache entries are used to route downlink packets (i.e., from gateway to MN) on the reverse path. Cache is refreshed also by *route-update packets* (empty IP packets) that are periodically sent to the gateway by MNs that are not regularly transmitting data. In this way the downlink routing state (*soft-state route*) can be maintained.

Handoffs are initiated by MNs on the basis of measurements of the BS signal strengths. While moving from BS3 to BS4 (see Fig. 49) during an active data session, the MN detects the stronger BS4 signal, tunes its radio to the channel used by BS4 and transmits a route-update packet (dotted line with "b" label in Fig. 49) that is cached by BSs along the path. BS2 adds to its routing cache the new mapping (MN, b), thus keeping a double entry related to MN (the old and the new route). Since the old mapping will be cleared only after the *routing-cache timeout* extinguishes, before this timeout both routes will coexist and packets addressed to MN will be delivered through both interfaces/path "a" and "b".

In the case that an MN does not receive packets for the *active-state-timeout,* it enters an *idle state*, letting its soft-state routing cache mappings time out. The following *paging* mechanism, derived from cellular telephony, is adopted by Cellular IP to reach idle hosts. *Paging-update packets* (i.e., empty IP datagrams) are periodically sent by the MN to the gateway in order to update the *paging cache* that is optionally maintained in Cellular IP nodes. When a node finds no valid routing cache mapping for an idle destination MN, *paging* occurs and IP packets are routed according to paging cache mappings (a node with no paging cache forward packets to all its interfaces except the source one). The paging cache mechanism allows avoiding broadcast search procedures.

Unlike in other solutions (e.g., HAWAII [166]), Cellular IP limits the use of explicit signaling messages and exploits IP datagrams for exchanging information on the position of mobile hosts. Moreover, it requires a simple configuration in the access network allowing easy employment and administration.

A 3G.IP group has been created to promote a common IP based wireless system for 3G mobile communication systems and to favor the standardization of an all IP-based wireless network architecture in 3GPP Releases 5 and 6 [167].

10.2   Wireless TCP

Most popular Internet applications, such as SMTP (e-mail), HTTP (WWW surfing) and FTP (file transfer), use the reliable services provided by TCP, a transport layer protocol in the Internet. The performance perceived by users mainly depend on the good behavior of TCP. Hence, studying its performance dynamics becomes a crucial part for the design of mobile networks that adopt the TCP/IP protocol suite.

TCP has been defined for traditional wired networks, characterized by low error rates and high bandwidth. The protocol interprets a packet loss in the network as an indication of network congestion (i.e., packet loss is due to the discard operated by a congested buffer encountered in the route), thus invoking congestion control and avoidance algorithms [168]. Such assumption is not correct over lossy links, such as wireless and satellite links, since packet losses are due to errors rather than to network congestion. Wireless links are characterized by low bandwidth, high latency, high bit error rates and temporary disconnections. In this environment the throughput at the TCP level may considerably degrade, thus affecting the behavior of applications.

Wireless networks share common characteristics, however, three main categories can be considered as different environments for data communications:

-   *Wireless Local Area Networks* (WLAN), with short links and high bandwidth;

-   *Wireless Wide Area Networks* (W-WAN), often referred to as *Long Thin Networks* (LTN), where "long" indicates high latency and "thin" stays for low bandwidth;

-   Satellite networks, often referred to as *Long Fat Networks* (LFN), where "fat" indicates high bandwidth.

The differences between them rely on the *Delay-Bandwidth Product* (DBP), that defines the capacity of a network path, that is the number of data segments that TCP should maintain "in flight" (i.e., sent but not yet acknowledged) in the channel in order to use efficiently the

available resources. Delay refers to the *Round Trip Time* (RTT), while
bandwidth refers to the capacity of the bottleneck in the network path.

Assuming for WLANs (of the IEEE 802.11 type) RTT = 3 ms and a
bandwidth of 1.5 Mbit/s, we obtain BDP = 4.5 Kbits. Instead, a 3G
cellular system (W-WAN) can offer a maximum bandwidth of 2 Mbit/s
and RTT = 200 ms, thus resulting in DBP = 50 Kbytes. This value is
higher than the standard dimension of a TCP buffer (8 Kbytes) adopted
by most TCP implementations; W-WANs will behave inefficiently
unless buffer dimension is incremented. Finally, a link between two
earth stations through a satellite GEO link presents a more critical
situation for channel efficiency and TCP performance [169]: assuming
RTT = 500 ms and a bandwidth of 36 Mbit/s, the result is DBP = 18
Mbits.

10.2.1 Mechanisms for improving wireless TCP performance on error-prone channels

When dealing with wireless links, two problems arise: one is due to the
characteristics of the link and the second is due to the mobility of the
receiver. Mobility can cause temporary disconnections due to handoffs
or to black holes in the coverage area. When disconnections are too
long, the sender could give up and close the TCP session. The
following three different approaches are possible for improving TCP
over wireless links:

- *End-to-end schemes*: they work at the transport layer, usually
  implementing the solution at the TCP sender;

- *Split-connection schemes*: it splits up the TCP connection by two, a
  wired connection (between the sender and the base station) and a
  wireless one (between the base station and the mobile terminal);

- *Link layer schemes*: these solutions do not directly affect TCP,
  since they are implemented in the link layer.

We limit the following study to a general level, without presenting the
detailed description of the specific protocols. Moreover, we assume that
the reader has a general background on TCP [170]-[172].

10.2.2 End-to-end approach

TCP grants reliability of data delivery by sending acknowledgments (ACK) from destination to source on an end-to-end basis. Optimization techniques at the transport layer are based on modifications to TCP only at the end points of a connection (see Fig. 50).

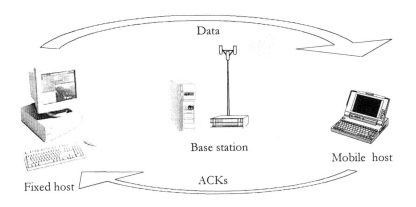

Fig. 50: End-to-end approach.

This approach does not alter the semantics of TCP sessions and it acts in a way to use more efficiently wireless links. Moreover, it should not affect the standard mechanisms for congestion control, like slow start and congestion avoidance.

An end-to-end scheme is the *Explicit Loss Notification* (ELN) [173] that adds an ELN option to TCP ACKs. After a packet loss in the wireless link, the future cumulative ACKs related to the lost packets are marked in order to signal that a non-congestion loss has occurred. Hence, the sender does not invoke any congestion control technique. This method changes TCP and does not solve the problem of temporary disconnections.

10.2.3 Split-connection approach

These solutions are based on the assumption that wired and wireless links have different characteristics and hence it is necessary to manage them separately (see Fig. 51). The neuralgic point of this approach is the *Performance Enhancing Proxy* (PEP), an intermediate node that allows to realize the two TCP connections and to exchange packets between them.

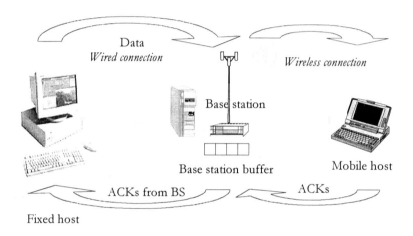

Fig. 51: Split connection approach.

The main advantage of this scheme is that congestion losses (in the wired links) and error losses (in the wireless links) can be separately treated and an appropriate wireless link specific protocol can be adopted for a better performance. However, TCP semantics is violated, since the fixed sender receives "false" ACKs, before data has successfully reached its final destination.

10.2.4 Link layer approach

The general idea of this approach is shown in Fig. 52. A link layer scheme is the *Snoop protocol* [174]. According to Snoop, PEP maintains a cache of TCP packets sent from the source and not yet acknowledged by the mobile. When Snoop detects a packet loss in the wireless link (either through a duplicate ACK or through a local timeout), it locally retransmits the packet. In this way, the Snoop protocol hides temporary degradations and occasional disconnections to the sender that does not invoke congestion control mechanisms.

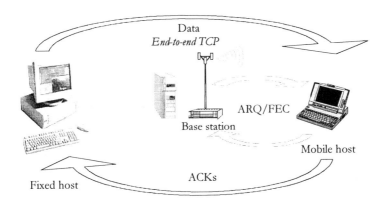

Fig. 52: Link layer approach.

The main disadvantage is the strong relation between the link layer that performs local retransmissions and the TCP layer. In fact, there is the possibility of having both the sender and the base station re-transmitting the same packet, especially in case of losses due to congestion. This fact can lead to bandwidth waste and performance degradation.

10.2.5 A final comparison

The best performance is achieved with a link layer protocol that is aware of TCP dynamics. However, end-to-end techniques seem to be more attractive, since significant improvements are obtained without any modification at intermediate nodes and without negative interference between different protocol layers.

References

[1] K. Pahlavan, "Wireless Communications for Office Information Networks", *IEEE Comm. Mag.*, Vol. 23, No. 6, pp. 19-27, June 1985.

[2] D. J. Goodman, "Cellular packet communications", *IEEE Trans. on Comm.*, Vol. 38, No. 8, pp. 1272 - 1280, August 1990.

[3] J. E. Padget, C. G. Gunther, T. Hattori, "Overview of Wireless Personal Communications", *IEEE Comm. Mag.*, Vol. 33, No. 1, pp. 28 - 41, January 1995.

[4] W. C. Y. Lee. *Mobile Cellular Telecommunication Systems.* McGraw-Hill, 1989.

[5] L. Kleinrock. *Queuing Systems* (Vol. I and II), J. Wiley & Sons, N.Y., 1976.

[6] W. C. Y. Lee, "Overview of cellular CDMA", *IEEE Trans. on Veh. Tech.*, Vol. 40, No. 2, pp. 291 - 302, May 1991.

[7] R. L. Pickholtz, L. B. Milstein, D. L. Schilling, "Spread Spectrum for Mobile Communications", *IEEE Trans. on Veh. Tech.*, Vol. 40, No. 2, pp. 313 - 322, May 1991.

[8] R. Prasad, T. Ojanpera, "An Overview of CDMA Evolution Toward Wideband CDMA", *IEEE Communications Surveys*, Vol. 1, pp. 2-29, Fourth Quarter 1998.

[9] J. Viterbi "Erlang Capacity of a Power Controlled CDMA System", *IEEE Journal Selected Areas in Communications*, Vol. 11, pp. 892-900, August 1993.

[10] J. G. Proakis, M. Salehi. *Communications Systems Engineering.* Prentice-Hall, Inc. New Jersey, 1994.

[11] M. Zorzi, "On the Analytical Computation of the Interference Statistics with Applications to the Performance Evaluation of Mobile Radio Systems", *IEEE Transactions on Communications*, Vol. 45, pp. 103-109, January 1997.

[12] F. Adachi, K. Ohno, A. Higashi, T. Dohi, Y. Okumura, "Coherent Multicode DS-CDMA Mobile Radio Access," *IEICE Trans. Commun.*, Vol. E79-B, No. 9, pp. 1316-1325, September 1996.

[13] Web site with URL: www.gsmworld.com.

[14] M. Mouly and M.-B. Pautet. *The GSM System for Mobile Communications.* 1992.

[15] ETSI, "Digital cellular telecommunications system (Phase 2+); General Packet Radio Service (GPRS); Service description"; Stage 1, GSM 02.60.

[16] 3GPP, "General Packet Radio Service (Release 1999); Service Description"; Stage 1, 3GPP TS 22.060, March 2000.

[17] G. Brasche, B. Walke, "Concepts, Services and Protocols of the new GSM; Phase 2+ General Packet Radio Service", *IEEE Comm. Mag.,* August 1997.

[18] J. Cai, D. J. Goodman, "General Packet Radio Service in GSM", *IEEE Comm. Mag.,* October 1997.

[19] C. Bettstetter, H.-J. Vögel, J. Eberspächer, "GSM Phase 2+ General Packet Radio Service Gprs: Architecture, Protocols, And Air Interface", *IEEE Communications Surveys,* Third Quarter 1999, Vol. 2, No. 3, pp. 2-14 (available at the Web address: http://www.comsoc.org/pubs/surveys/3q99issue/).

[20] R. Kalden, I. Meirick, M. Meyer, "Wireless Internet Access Based on GPRS", *IEEE Personal Comm.,* Vol. 7, No. 1, April 2000.

[21] A. S. Tanenbaum. *Computer Networks.* Prentice-Hall International, Inc., New Jersey, 1996.

[22] ETSI, "Overall description of GPRS radio interface", GSM 03.64, ver. 6.2.0, 5-1999.

[23] ETSI, "Digital cellular telecommunications system (Phase 2+); Channel coding", GSM 05.03 1998.

[24] ETSI, "Digital cellular telecommunications system (Phase 2+); General Packet Radio Service (GPRS); Service Description"; Stage 2, GSM 03.60.

[25] ETSI, "Digital cellular telecommunications system (Phase 2+); General Packet Radio Service (GPRS); Mobile Station (MS) - Base Station System (BSS) interface; Radio Link Control / Medium Access Control (RLC/MAC) protocol", GSM 04.60.

[26] P. Stuckmann, "Quality of Service Management in GPRS-based Radio Access Networks", *Telecommunication Systems,* Vol. 19, No. 3-4, pp. 515-546, January 2002.

[27]   3GPP, "General Packet Radio Service (Release 1999); Service Description"; stage 2, 3G TS 23.060, January 2000.

[28]   ETSI, "Digital cellular telecommunications system (Phase 2+); General Packet Radio Service (GPRS); GPRS Tunnelling Protocol (GTP) across the Gn and Gp Interface", ETSI GSM 09.60.

[29]   3GPP, "General Packet Radio Service (Release 1999); GPRS Tunnelling Protocol (GTP) across the Gn and Gp Interface", 3G TS 29.060, January 2000.

[30]   ETSI, "General Packet Radio Service (GPRS); Mobile Station (MS) - Serving GPRS Support - Logical Link Control (LLC) layer specification", Phase 2+, GSM 04.64 - TS 101 351 V7.0.0, Release 1998.

[31]   ETSI "General Packet Radio Service (GPRS); Mobile Station (MS) - Serving GPRS Support Node (SGSN); Subnetwork Dependent Convergence Protocol (SNDCP)", Phase 2+, GSM 04.65 - TS 101 297, V6.4.0, Release 1997.

[32]   ETSI, "General Packet Radio Service (GPRS); Base Station System (BSS) - Serving GPRS Support Node (SGSN); BSS GPRS Protocol (BSSGP)", Phase 2+, GSM 08.18 - ETSI TS 101 343, V6.4.0, Release 1997.

[33]   R. Prasad, W. Mohr, W. Kon. *Third Generation Mobile Communication System*. Artech House, London, 2000.

[34]   UMTS Forum Web site with URL: http://www.umts-forum.org

[35]   ATDMA   RACE   II   Project   with   URL: http://www.cordis.lu/infowin/acts/analysys/concertation/mobilit y/racedel.html

[36]   3GPP Web site with URL: http://www.3gpp.org/

[37]   3GPP2 Web site with URL: http://www.3gpp2.org/

[38]   N. Dimitriou, G. Sfikas, R. Tafazolli, "Quality of Service for Multimedia CDMA", *IEEE Comm. Mag.*, Vol. 38, No. 7, July 2000.

[39]   3GPP, "QoS Concept and Architecture", TS 23.107 v3.3.0, June 2000.

[40]   3GPP, "Network architecture", TS 23.002.

[41]   3GPP "Universal Mobile Telecommunications System (UMTS); UTRAN Iu interface data transport and transport signalling", TS 25.414 version 3.9.0 Release 1999.

[42]   ITU, "B-ISDN ATM Layer (AAL) Functional Description", I.362.

[43]   ETSI, "Evolution of the GSM platform towards UMTS", UMTS 23.20 version 1.0.0, 1998.

[44]   3GPP, "General aspects & Principles of Iu interface between CN and RAN", TS 25.410.

[45]   IETF RFC 2474, "Definition of the Differentiated Services Field (DS Field) in the IPv4 and IPv6 Headers", December 1998 (http://www.ietf.org/rfc/).

[46]   IETF RFC 2475, "An Architecture for Differentiated Services", December 1998 (http://www.ietf.org/rfc/).

[47]   IETF RFC 2460, "Internet Protocol, Version 6 (IPv6) Specification", December 1998 (http://www.ietf.org/rfc/).

[48]   IPv6 forum Web site with address: http://www.ipv6forum.org/

[49]   IETF RFC 791, "INTERNET PROTOCOL, DARPA INTERNET PROGRAM, PROTOCOL SPECIFICATION", September 1981 (http://www.ietf.org/rfc/).

[50]   3GPP, "Packet switched conversational multimedia applications; Transport protocols", TS 26.236 Release 5.

[51]   IETF RFC 2543, "SIP: Session Initiation Protocol", March 1999 (http://www.ietf.org/rfc/).

[52]   ETSI, "Digital cellular telecommunications system (Phase 2+); Universal Mobile Telecommunications System (UMTS); End to end quality of service concept and architecture", (3GPP TS 23.207 version 5.4.0 Release 5)

[53]   ETSI, "Universal Mobile Telecommunications System (UMTS); *Selection procedures for the choice of radio transmission technologies of the UMTS*" (UMTS 30.03 version 3.2.0), ETSI TR 101 112 V3.2.0 (1998-04).

[54]   3GPP, "UE radio transmission and reception (FDD)", TS 25.101.

[55]   3GPP, "UE radio transmission and reception (TDD)", TS 25.102.

[56] 3GPP, "RF parameters in support of Radio Resource Management", TS 25.103.

[57] 3GPP, "BTS radio transmission and reception (FDD)", TS 25.104.

[58] 3GPP, "BTS radio transmission and reception (TDD)", TS 25.105.

[59] 3GPP, "Base station conformance testing (FDD)", TS 25.141.

[60] 3GPP, "Base station conformance testing (TDD)", TS 25.142.

[61] 3GPP, "Base station EMC", TS 25.143.

[62] 3GPP, "Physical layer general description", TS 25.201.

[63] 3GPP, "Physical Channels and Mapping of Transport Channels onto Physical Channels (FDD)", TS 25.211.

[64] 3GPP, "Multiplexing and channel coding (FDD)", TS 25.212.

[65] 3GPP, "Spreading and modulation (FDD)", TS 25.213.

[66] 3GPP, "Physical layer procedures (FDD)", TS 25.214.

[67] 3GPP, "Transport channels and physical channels (TDD)", TS 25.221.

[68] 3GPP, "Multiplexing and channel coding (TDD)", TS 25.222.

[69] 3GPP, "Spreading and modulation (TDD)", TS 25.223.

[70] 3GPP, "Physical layer procedures (TDD)", TS 25.224.

[71] 3GPP, "Physical layer - measurements", TS 25.231.

[72] 3GPP, "Radio Interface Protocol Architecture", TS 25.301.

[73] 3GPP, "Services Provided by the Physical Layer", TS 25.302.

[74] 3GPP, "UE Functions and Inter-layer procedures in Connected Mode", TS 25.303.

[75] 3GPP, "UE Functions Related to Idle Mode", TS 25.304.

[76] 3GPP, "Medium Access Control (MAC) Protocol Specification", TS 25.321.

[77] 3GPP, "Radio Link Control (RLC) Protocol Specification", TS 25.322.

[78] 3GPP, "Radio Resource Control (RRC) Protocol Specification", TS 25.331.

[79]   3GPP, "Channel coding and multiplexing examples", TR 25.944.

[80]   3GPP "The Virtual Home Environment", TS 22.121 V5.2.0 (2001-12).

[81]   G. Maral and M. Bousquet. *Satellite Communications Systems.* 3rd ed. New York, NY: John Wiley and Sons, 1998.

[82]   Ufficial Globalstar Web-site: http://www.globalstar.com

[83]   Ufficial Teledesic Web-site: http://www.teledesic.com

[84]   Ufficial   Skybridge   Web-site: http://www.skybridgesatellite.com/

[85]   Ufficial   Spaceway   Web-site: http://www.skybridgesatellite.com/

[86]   Ufficial DirectPC Web site with URL: http://www.directpc.com

[87]   Ufficial Starband Web site with URL: http://www.starband.com

[88]   E. Del Re, "A Coordinated European Effort for the Definition of a Satellite Integrated Environment for Future Mobile Communications", *IEEE Comm. Mag.*, Vol. 34, No. 2, pp. 98-104, February 1996.

[89]   Y. F. Hu, E. Sheriff, E. Del Re, R. Fantacci, G. Giambene, "Satellite-UMTS Dimensioning and Resource Management Technique Analysis", *IEEE Trans. on Veh. Tech.*, Vol. 47, No. 4, pp. 1329-1341, November 1998.

[90]   3GPP, "Technical Specification Group Services and System Aspects, Iu Principles", 3G TR 23.930.

[91]   P. Taaghol, B. G. Evans, E. Buracchini, R. De Gaudenzi, G. Gallinaro, J. Ho Lee, C. Gu Kang, "Satellite UMTS/IMT2000 W-CDMA Air Interfaces", *IEEE Comm. Mag.,* Vol. 37, No. 9, pp. 116-126, September 1999.

[92]   ETSI, S-UMTS-A 25.211 - 25.214 specifications on physical layer aspects, 2000.

[93]   D. J. Bem, T. W. Wieckowski, R. J. Zielinsky, "Broadband Satellite Systems", *IEEE Comm. Surveys*, Vol. 3, No. 1, 2000; available at the URL: http://www.comsoc.org/pubs/surveys

[94]   J. Neale, R. Green, J. Landovskis, "Interactive Channel for Multimedia Satellite Networks", *IEEE Comm. Mag.*, Vol. 39, No. 3, pp. 2-8, March 2001.

[95] J. Farserotu, R. Prasad, "A Survey of Future Broadband Multimedia Satellite Systems, Issues and Trends", *IEEE Comm. Mag.*, June 2000.

[96] S. Ohmori, Y. Yamao, N. Nakajima, "The Future Generations of Mobile Communications Based on Broadband Access Technologies", *IEEE Comm. Mag.*, December 2000.

[97] R. Ramjee, T. F. La Porta, L. Salgarelli, S. Thuel, K. Varadhan, L. Li, "IP-Based Access Network Infrastructure for Next-Generation Wireless Data Networks", *IEEE Personal Comm.*, Vol. 7, No. 4, pp. 34- 41, August 2000.

[98] M. Dinis, J. Fernandes, "Provision of Sufficient Transmission Capacity for Broadband Mobile Multimedia: A Step Toward 4G", *IEEE Comm. Mag.*, pp. 46-54, August 2001.

[99] Web site with URL: http://users.ece.gatech.edu/~jxie/4G/

[100] OFDM forum with URL: www.ofdm-forum.com.

[101] HIPERLAN/2 Web site with URL: www.hiperlan2.com.

[102] M. de Lurdes Lourenco, "The advantage of MBS over wireless LAN's: comparison of MBS with Hiperlan", *Proc. of RACE Mobile Workshop*, Amsterdam (NL), pp. 519-524, May 17-19, 1994.

[103] *Mobile Broadband Systems* (MBS) project home page at ULR: http://www.comnets.rwth-aachen.de/project/mbs/.

[104] D. Petras, A. Krämling, "Wireless ATM Performance evaluation of a DSA++ MAC protocol with fast collision resolution by a probing algorithm", *Int. J. of Wireless Information Networks*, Vol. 4, No. 4, October 1997.

[105] N. Passas, L. Merakos, D. Skyrianoglou, F. Bauchot, S. Decrauzat, "MAC Protocol and Traffic Scheduling for Wireless ATM Networks", *Mobile Networks and Applications*, pp. 275-292, 1998.

[106] EU projects Web site with URL: www.ideal-ist.net/

[107] G. J. Miller, K. Thompson, R. Wilder, "Wide-Area Internet Traffic Patterns and Characteristics", *IEEE Network*, pp. 10-23, November/December 1997.

[108] A. Chandra, V. Gumalla, J. O. Limb, "Wireless Medium Access Control Protocols", *IEEE Communications Surveys*, Second

Quarter   2000   (available   at
http://www.comsoc.org/pubs/surveys).

[109]   S. Nanda, D. J. Goodman, U. Timor, "Performance of PRMA: a
        Packet Voice Protocol for Cellular Systems", *IEEE Trans. On
        Veh. Tech.*, Vol. 40, No. 3, pp. 584-598, August 1991.

[110]   C. Cleary, M. Paterakis, "Design and Performance Evaluation
        of a Scheme for Voice-data Channel Access in Third
        Generation Microcellular Wireless Networks", *Mobile Networks
        and Applications (MONET) Journal*, ACM and Baltzer Science
        Publishing, Vol. 2, No. 1, pp. 31-43, January 1997.

[111]   R. Guerin, V. Peris, "Quality-of-service in Packet Networks:
        basic Mechanisms and Directions", *Computer Networks*, Vol.
        31, pp. 169-189, 1999.

[112]   K. Dimyati, Y. T. Chin, "Policing Mechanism and Cell Loss
        Priority Control on Voice Cells in ATM Networks Using Fuzzy
        Logic", *IEEE Proc.-Comm.*, Vol. 147, No. 4, August 2000.

[113]   3GPP, "Technical Specification Group Radio Access Network;
        Improvement of RRM across RNS and RNS/BSS", 3GPP TR
        25.881, Release 5, 2001.

[114]   C. Blondia, O. Casals, "Performance Analysis of Statistical
        Multiplexing of VBR sources", *Proc. of INFOCOM'92*, pp.
        828-838.

[115]   B. Maglaris, D. Anastasiou, Prodip Sen, G. Karlsson, J. D.
        Robbins, "Performance Models of Statistical Multiplexing in
        Packet Video Communications", *IEEE Trans. on Comm.*, Vol.
        36, No. 7, pp. 834-843, July 1988.

[116]   "WAND design requirements", CEC Deliverable, n. 1D3,
        available at the URL: http://www.tik.ee.ethz.ch/.

[117]   A. E. Brand, A. H. Aghvami, "Multidimensional PRMA with
        Prioritized Bayesian Broadcast – A MAC Strategy for
        Multiservice Traffic over UMTS", *IEEE Trans. on Veh. Tech.*,
        Vol. 47, No. 4, pp. 1148-1161, November 1998.

[118]   A. Andreadis, G. Benelli, G. Giambene, B. Marzucchi,
        "Analysis of the WAP Protocol over SMS in GSM Networks",
        Wireless Communications and Mobile Computing Journal, pp.
        381-395, Vol. 1, No. 4, October-December 2001.

[119] M. E. Crovella, A. Bestavros: "Self-Similarity in World Wide Web Traffic: Evidence and Possible Causes", *IEEE/ACM Transactions on Networking*, Vol. 5, No. 6, pp. 835-846, December 1997.

[120] I. Norros, "A Storage Model with Self-similar Input", *Queueing Systems*, Vol. 16, pp. 387-396, 1994.

[121] R. G. Addie, M. Zukerman, T. D. Neame, "Broadband Traffic Modeling: Simple Solution to Hard Problems", *IEEE Comm. Mag.*, pp. 89-95, August 1998.

[122] W. C. Jakes. *Microwave Mobile Communications*, New York: Wiley, 1974.

[123] E. N. Gilbert, "Capacity of Burst Noise Channels", *Bell Sys. Tech. J.*, Vol. 39, pp. 1253-1256, 1960.

[124] A. A. Hassan, S. Channakeshu, J. B. Andreson, "Performance of Coded Slow-Frequency-Hopped TDMA Cellular Systems", *Proc. of IEEE Veh. Tech. Conf.*, pp. 289-292, 1993.

[125] Michele Zorzi, Ramesh R. Rao, Laurence B. Milstein, "Error Statistics in Data Transmission over Fading Channels", *IEEE Trans. Comm.*, Vol. 46, No. 11, pp.1468-1477, November 1998.

[126] *GPRS From A-Z*, Edited by INACON GmbH Training and Consulting.

[127] Y. Lu, R. Brodersen, "Unified Power Control, Error Correction Coding and Scheduling for a CDMA Downlink System", *Proc. of IEEE INFOCOM '96*, Vol. 3, pp. 1125-1132.

[128] Ö. Gürbüz, H. Owen, "Dynamic Resource Scheduling Strategies for QoS in W-CDMA", *Proc. of IEEE GLOBECOM '99*, pp. 183-187.

[129] Ö. Gürbüz, H. Owen, "Dynamic Resource Scheduling Schemes for W-CDMA Systems", *IEEE Comm. Mag.*, pp. 80-84, October 2000.

[130] Ö. Gürbüz, H. Owen, "Dynamic Resource Scheduling for Variable QoS Traffic in W-CDMA", *Proc. of IEEE ICC '99*, Vol. 2, pp. 703-707, June 1999.

[131] Ö. Gürbüz, H. Owen, "A Resource Management Framework for QoS Provisioning in W-CDMA Systems", *Proc. of IEEE VTC '99*, Vol. 1, pp. 407-411, May 1999.

[132] A. Sampath, P. S. Kumar, P. Holtzman, "Power Control and Resource Management for a Multimedia CDMA Wireless System", *Proc. of PIMRC'95*, Vol. 1, pp. 21-25, September 1995.

[133] 3GPP, "Radio resource management strategies", TS 25.922.

[134] K. Balachandran, K. Chang, W. Luo, S. Nanda, "System Level Interference Mitigation Schemes in EGPRS: Mode-0 and Scheduling", *Proc. of VTC '01*, Spring 2001, pp. 2489-2493.

[135] 3GPP, "Mobile radio interface signalling layer 3; General aspect", TS 24.007 (Release 4) v4.0.0, April 2001.

[136] M. Haardt, A. Klein, R. Koehn, S. Oestreich, M. Purat, Volker Sommer, T. Ulrich, "The TD-CDMA Based UTRA TDD Mode", *IEEE Journal on Sel. Areas in Comm.*, Vol. 18, No. 8, August 2000.

[137] 3GPP, "Technical Specification Group (TSG) RAN WG4; RF System Scenarios", TR 25.942 V2.2.1 (1999-12).

[138] C.-S. Chang, K.-C. Chen, M.-Y. You, J.-F- Chang, "Guaranteed Quality-of-Service Wirelss Access to ATM Networks", *IEEE Journal on Sel. Areas Comm.*, Vol. 15, No. 1, pp. 106-118, 1997.

[139] A. S. Mahmoud, D. D. Falconer, S. A. Mahmoud, "A Multiple Access Scheme for Wireless Access to a Broadband ATM LAN Based on Polling and Sectored Antenna", *IEEE Journal on Sel. Areas in Comm.*, Vol. 14, No. 4, pp. 596-608, 1996.

[140] Andreadis, G. Benelli, G. Giambene, F. Partini, "Integration of Video and Data Bursty Traffics in Wireless ATM Networks", *Wireless Personal Communications journal - Kluwer Academic Publishers*, Vol. 19, pp. 243-262, 2001.

[141] G. Benelli, R. Fantacci, G. Giambene, C. Ortolani, "Performance analysis of a PRMA protocol suitable for voice and data transmissions in low earth orbit mobile satellite systems", *IEEE Transactions on Wireless Communications*, Vol. 1, No. 1, pp. 156 -168, January 2002.

[142] R. Fantacci, G. Giambene, R. Angioloni, "A Modified PRMA Protocol for Voice and Data Transmissions in Low Earth Orbit Mobile Satellite Systems", *IEEE Trans. on Veh. Tech.*, Vol. 49, No. 5, pp. 1856-1876, September 2000.

[143] G. Giambene, "MAC Schemes for LEO Satellites", Lecture of *the first Online Symposium for Electronics Engineers in USA*, October 10, 2000, Lecture available on line at the following address: www.techonline.com.

[144] A. Fukuda, S. Tasaka, "The Equilibrium Point Analysis - A Unified Analytic Tool for Packet Broadcast Networks", *Proc. of IEEE GLOBECOM'83*, San Diego, CA, November 1983, pp. 33.4.1 - 33.4.8.

[145] S. Tasaka, "Performance Comparison of Multiple Access Protocols for Satellite Broadcast Channels", *Proc. of IEEE GLOBECOM'83*, San Diego, CA, November 1983, pp. 35.3.1 - 35.3.8.

[146] S. Nanda, "Stability Evaluation and Design of the PRMA Joint Voice Data system", *IEEE Trans. Comm.*, Vol. 42, pp. 2092 - 2104, May 1994.

[147] P. T. Saunders. *An Introduction to Catastrophe Theory.* Cambridge University press, NY, USA, 1980.

[148] M. Golubitsky "An Introduction to Catastrophe Theory and its Applications", *SIAM Review*, Vol. 20, No. 2, pp. 352-387, 1978.

[149] B. Steyaert, H. Bruneel, G. H. Petit, D. De Vleeschauwer, "A Versatile Queueing Model Applicable in IP Traffic Studies", *COST 257 Project*, TD (00)-02, Barcelona, January 2000.

[150] F. Harvey, "The Internet in Your Hand", *Scientific American*, October 2000.

[151] K. J. Bannan, "The Promise and Perils of WAP", *Scientific American*, October 2000.

[152] WAP Forum Web site with URL: http://www.wapforum.org/.

[153] B. Hu, "Wireless Portal Technology - An Overview and Perspective", *Proc. of the First on Line Symposium for Electrical Engineers*, October 2000.

[154] WAP Forum, "Wireless Application Protocol, WAP 2.0 Technical White Paper", *paper available at the WAP Forum Web site* with URL: http://www.wapforum.org/.

[155] Example of WAP gateway characteristics: Nokia WAP gateway, available at the URL: http://www.nokia.com/corporate/wap/gateway.html.

[156] Wireless Application Protocol Forum, Ltd., "Binary XML Content Format Specification", WAP-154 Version 1.2, November 4, 1999.

[157] Sungwon Lee, Nah-Oak Song, "Experimental WAP (Wireless Application Protocol) Traffic Modeling on CDMA based Mobile Wireless Network", *Proc. of the 54th Veh. Tech. Conf., 2001*, VTC 2001 Fall, pp. 2206-2210, 2001.

[158] D. Ralph, H. Aghvami, "Wireless Application Protocol Overview", *Wireless Communications and Mobile Computing Journal*, pp. 125-140, Vol. 1, No. 2, April-June 2001.

[159] Wireless Application Protocol Forum, Ltd., "Wireless Datagram Protocol Specification", WAP-158, November 5, 1999.

[160] WAP browsers: Nokia from http://www.nokia.com; Ericsson from http://www.ericsson.se/WAP; UP.Browser from Phone.com from http://updev.phone.com; WinWAP from http://www.slobtrot.com; Motorola from http://www.motorola.com; Gelon.net from http://www.gelon.net; WAPman from http://palmsoftware.tucows.com.

[161] 3GPP, "Technical Specification Group Terminals; Mobile Station Application Execution Environment (MExE); Functional description". Stage 2, 3G TS 23.057.

[162] IETF, "IP Mobility Support", RFC 2002, October 1996 (http://www.ietf.org/rfc/).

[163] P. Reinbold, O. Bonaventure, "A Comparison of IP Mobility Protocols", *IEEE SCVT 2001 Proceedings*, June 2001.

[164] A. G. Valko, "Cellular IP: A New Approach to Internet Host Mobility", *ACM Computer Communication Review*, January 1999.

[165] A. T. Campbell, J. Gomez, S. Kim, A. G. Valko and C. Wan, "Design, Implementation, and Evaluation of Cellular IP", *IEEE Personal Communications*, August 2000.

[166] R. Ramjee et al., "IP micro-mobility support using HAWAII", *Internet Draft*, draft-ietf-mobileip-hawaii-01.txt, July 2000.

[167] 3G.IP Web site with URL: http://www.3gip.org/

[168] V. Jacobson, "Congestion Avoidance and Control", *Computer Communications Review*, Vol. 18, No. 4, pp. 314-329, August 1988.

[169] IETF RFC 2488, "Enhancing TCP Over Satellite Channels Using Standard Mechanisms ", January 1999.

[170] IETF RFC 793, "Transmission Control Protocol ", September 1981.

[171] D. E. Comer. *Internetworking with TCP/IP, Vol.1: Principles, Protocols and Architecture*. Prentice-Hall, 1995.

[172] W. Stevens. *TCP/IP Illustrated, Volume 1; the Protocols*. Addison Wesley, 1995.

[173] H. Balakrishnan, V. N. Padmanabhan, S. Seshan, R. H. Ketz, "A Comparison of Mechanisms for Improving TCP Performance over Wireless Links", *Proc. of ACM SIGCOMM'96*, Stanford, CA, August 1996.

[174] H. Balakrishnan, S. Seshan, E. Amir, R. Katz, "Improving TCP/IP Performance over Wireless Networks", *Proc. of ACM Mobicom,* Berkeley, CA, December 1995.

Book index